Statement of John F. Tierney
Chairman
Subcommittee on National Security and Foreign Affairs
Committee on Oversight and Government Reform
U.S. House of Representatives

Hearing on "Rise of the Drones: Unmanned Systems and the Future of War"

<u>**As Prepared for Delivery**</u>

March 23, 2010

Good morning. Our hearing today introduces a new topic to the Subcommittee: the rise of unmanned systems, and their implications for U.S. national security.

Over the last decade, the number of unmanned systems and their applications has grown rapidly. So too has the number of operational, political, and legal questions associated with this technology.

The growing demand for and reliance on unmanned systems has serious implications, both on and off the battlefield. As the United States is engaged in two wars abroad, unmanned systems, particularly unmanned aerial vehicles, have become a centerpiece of that war effort. In recent years, the Department of Defense's UAV inventory has rapidly grown in size, from 167 in 2002 to over seven thousand today. Last year, for the first time, the U.S. Air Force trained more unmanned pilots than traditional fighter pilots.

Some express no doubt that unmanned systems have been a boost to U.S. war efforts in the Middle East and South Asia. CIA Director Leon Panetta said last May that "drone strikes are the only game in town in terms of confronting or trying to disrupt the al Qaeda leadership." Media reports over the last year that the top two leaders of the Pakistani Taliban were killed by drone strikes also support this argument.

But some critics argue that drone strikes are unethical at best and counter-productive at worst. They point to the reportedly high rate of civilian casualties, which has been calculated by the *New American Foundation* to be around 32 percent, and argue that the strikes do more to stoke anti-Americanism than they do to weaken our enemies. A quick skim of any Pakistani newspaper provides some evidence to support this theory. This is particularly relevant in the era of counter-insurgency doctrine, a central tenet of which is, 'first, do no harm.'

It also may be the case that we are fighting wars with modern technology under an antiquated set of laws. For example, if the United States uses unmanned weapons systems, does that require an official declaration of war or an authorization for the use of force?

Do the Geneva Conventions – written in 1949 – govern the prosecution of an unmanned war?

Who is considered a lawful combatant in unmanned war – the Air Force pilot flying a Predator from thousands of miles away in Nevada, or the civilian contractor servicing it in on an airstrip in Afghanistan?

If unmanned systems are changing the way that we train our military personnel, so too should they change the way that we respond to the stress of combat. We already know that unmanned pilots are showing signs of equal or greater stress from combat compared to traditional pilots. The stress of fighting a war thousands of miles away, then minutes later joining your family at the dinner table presents mental health challenges that we must address.

On the domestic front, manufacturers have already developed a number of unmanned commercial products, and are likely to find more applications for this technology in the future. From vacuum cleaners to crop dusters, traditional items that require manual operation are rapidly being rendered obsolete by unmanned technology. UAVs are now being used for environmental monitoring, particularly in hard to reach places like the North Pole. Last fall, the University of North Dakota chartered a four-year degree program in UAV piloting.

These trends are already forcing us to ask new questions about domestic airspace regulation: who is allowed to own unmanned systems, and where they are allowed to operate them? Additionally, as more law enforcement and border security services come to use unmanned systems, important questions continue to emerge about the protection of privacy.

As this technology develops and becomes more commercially available, we must implement adequate measures to prevent it from falling into the wrong hands. At least 40 other countries are currently developed unmanned systems technology – including Iran, Russia, and China.

We already know that during the Israel-Lebanon war in 2006, Hezbollah deployed three surveillance UAVs that it acquired from Iran. A recent Air Force study concluded that a UAV is an ideal platform for a chemical or biological terrorist attack. As Peter Singer, one of our witnesses today, wrote recently in *Newsweek*, "for less than $50,000, a few amateurs could shut down Manhattan."

We must ensure that the appropriate government agencies are coordinating their efforts to prevent this technology from proliferating and falling into the wrong hands, and also to ensure that we have adequate homeland security measures to respond to these threats.

Finally, as this new technology continues to develop, we must ensure that there are adequate measures in place to prevent waste, fraud, and abuse in the acquisition process. A 2009 study by the U.S. Government Accountability Office, the author of which we will hear from today, reported significant cost growth, schedule delays, and performance shortfalls in DOD's UAV acquisition process. This analysis raises serious concerns, and I look forward to learning more on this from both the GAO and DOD witnesses before us.

These are some of the questions that we will begin to answer in this hearing. Surely we will not conclude this conversation in one afternoon, but I hope that this hearing serves as a thorough introduction to the topic, for the purpose of educating and informing our Members, as well as the American public.

Testimony by Kevin J. Wolf
Assistant Secretary for Export Administration
Bureau of Industry and Security
U. S. Department of Commerce

Before the National Security and Foreign Affairs
Subcommittee of the Committee on Oversight and Reform

March 23, 2010

Chairman Tierney, Congressman Flake, Members of the Committee:

Thank you for the opportunity to testify before the National Security and Foreign Affairs Subcommittee of the Committee on Oversight and Reform, on the Department's role in export controls of unmanned aerial vehicles (UAV) and related technology.

The Bureau of Industry and Security (BIS) within the Department of Commerce, administers controls on the export of a range of dual-use items (commodities, software, and technology with civilian and military uses). In doing so, BIS works closely with a number of departments and agencies, including the Departments of Defense, State, and Energy, the Central Intelligence Agency, the Department of Homeland Security's Bureau of Immigration and Customs Enforcement, and the Department of Justice.

The dual-use export control system is an important tool to protect the national security of the United States against the diverse threats our nation faces. State and non-state actors seek to acquire weapons of mass destruction (WMD) and the means to deliver them, as well as conventional arms and other items that could be used for terrorist purposes. BIS implements the dual-use control system through the Export Administration Regulations (EAR). Under the EAR, BIS regulates the export of certain UAVs and related items based on multilateral control lists and other items that could be used in or for UAVs through unilateral controls on end-uses and end-users.

U.S. Dual-Use Export Controls Relating to UAVs

A. The Missile Technology Control Regime

The United States has a comprehensive export control program, consisting of multilateral commitments and unilateral controls, intended to prevent the proliferation of sensitive items to countries and programs of concern. The Missile Technology Control Regime (MTCR) has 34 member countries, including many of the key manufacturers and exporters of cruise missiles and UAVs. The MTCR has a control list, or "Annex," of items (goods and technologies) which all members control according to the MTCR guidelines – including certain UAVs. The MTCR Guidelines and Annex serves as the basis for the dual-use missile technology controls set forth in the EAR. The MTCR definition of UAVs includes cruise missiles, target drones, reconnaissance drones, and other forms of UAVs regardless of whether they are military or civilian or armed or unarmed. UAVs can be as large as a commercial airplane or as small as a model airplane and jet or propeller driven; they can be airplanes, helicopters, or even blimps; they can be guided autonomously or by a remote operator or pilot. But only UAVs meeting certain range/payload criteria are MTCR-controlled.

The MTCR also addresses newly emerging technologies, the application of new uses for old items, and requests for the imposition of additional controls. The Department of Commerce, along with the Departments of State, Defense, and Energy, actively participates in the interagency Missile Annex Review Committee (MARC). The MARC is responsible for reviewing internal and foreign proposals for modifying existing MTCR control parameters or assessing proposals for new MTCR controls.

The MTCR, from its inception in 1987, subjected exports of UAVs capable of delivering a payload of at least 500 kg to a range of at least 300 km (so-called "Category I" or "MTCR-class" UAVs) and their directly associated technology to a "strong presumption of denial." Exports of complete warhead safing, arming, fuzing, firing subsystems useable in such UAVs, and their directly associated technology, also are subject to a "strong presumption of denial." In addition, under the MTCR Guidelines, the transfer of MTCR-defined "production facilities" for Category I items and the technology directly associated with these facilities will not be authorized.

Key components and materials useable in producing MTCR-class UAVs -- such as small, fuel-efficient jet engines; structural composites and their production equipment; various types of avionics, guidance, and flight control systems; telemetry and ground support equipment; various test equipment; and stealth technology -- are controlled as MTCR Category II items. MTCR countries review exports of such items on a case-by-case basis against specified nonproliferation criteria, and such exports also are subject to a "strong presumption of denial" if assessed to be intended for use in WMD delivery. In 1994, additional UAVs -- those not captured under Category I, but inherently capable of a 300 km range regardless of payload capability -- were added to Category II MTCR controls.

In addition to MTCR controls, UAVs and their components are controlled under the Wassenaar Arrangement, the multilateral export control regime for conventional arms and associated dual-use items. The Department of Commerce implements Wassenaar-related controls over non-military UAVs having either: a) an autonomous flight control and navigation capability (e.g., an autopilot with an inertial navigation system); or b) the capability of controlled-flight out of direct vision range involving a human operator (e.g., televisual remote control). Wassenaar also requires controls on the export of a wide range of materials and equipment that may be used in the production of UAVs, beyond those controlled by the MTCR.

Moreover, there are a large number of UAV-relevant items that are not controlled under the MTCR or Wassenaar, mostly because of their broad civil uses (e.g., in manned aircraft). On a national basis, the U.S. and most other members of the multilateral export control regimes have enacted "catch-all" controls to cover exports of such items when an exporter knows or is informed by his government that they are intended for use in WMD programs (including WMD delivery).

Thus, continuing to work within the multilateral MTCR framework is essential to the success of our missile and UAV nonproliferation goals.

B. U.S. Implementation of Missile Technology Export Controls

Consistent with its MTCR commitments, the United States implements a comprehensive export control program intended to prevent the proliferation of sensitive items to programs of concern. The Department of State has export licensing jurisdiction for defense articles and services covered by the U.S. Munitions List, including all military UAVs regardless of range or payload, commercial UAVs with a range of at least 300km and a payload capability of 500 kg as well as certain related components and technologies. The Department of Commerce has export licensing jurisdiction for dual-use items (items with civilian and military applications) enumerated on the Commerce Control List (CCL), as well as items not on the CCL but subject to the EAR. The Department of Commerce also has jurisdiction over certain WMD and missile-related activities of U.S. persons.

The Department of Commerce uses a number of tools to prevent the proliferation of items under its jurisdiction related to cruise missiles and UAVs. First, the CCL contains a list of items controlled for Missile Technology (MT) reasons. These MT-controlled items encompass the equipment and technologies that the MTCR has agreed are of proliferation concern and not already controlled as munitions items.

Under the EAR, an exporter must submit a license application to export any item controlled for MT reasons to any country in the world (except Canada). Since January 2003, BIS has issued 96 licenses for items covered by Export Control Classification Number (ECCN) 9A012 on the Commerce Control List which is a part of the EAR. ECCN 9A012 covers certain non-military UAVs, associated systems, equipment, and components. Twenty-five of these licenses have actually been used to export UAV and UAV components totaling just over $4 million dollars. Of the 25 exports the vast majority went to MTCR Partner Countries in Europe, Australia, South America, and South Korea with the remaining going to Mexico, Singapore, Indonesia and Iraq. The Departments of Defense, State, and Energy, as well as Commerce, review all license applications for MT-controlled items. The reviewing departments apply the MTCR Guidelines and additional criteria, consider available intelligence and law enforcement information, and determine if the transaction would pose an unacceptable risk of diversion or provide a material contribution to a missile program of concern.

In addition, the interagency Missile Technology Export Control (MTEC) group meets once a week to review all pending missile technology license applications. The process for interagency review of export license applications submitted to the Department of Commerce established by Executive Order 12981, as amended, ensures the positions of the reviewing departments are fully considered before an export license is approved.

The U.S. controls on exports that could support WMD and missile programs go well beyond the MTCR Annex items. Under our catch-all controls, exporters also are required to obtain a license for the export, reexport, or in-country transfer of any item, even a non-controlled item, if they know or are informed that the item will be used in or for prohibited nuclear activities, chemical or biological weapons programs, or the design, development, or production of missiles, or by facilities engaged in such activities. The definition of missile includes UAVs with a range equal to or greater than 300 kilometers.

These catch-all controls, set forth in Part 744 of the EAR, seek to prevent the export, reexport, or in-country transfer of any item that could be used in a missile program of concern, and thus specifically detail the inclusion of cruise missiles and UAVs capable of performing military reconnaissance, surveillance, or combat support to ensure there is no "gap" in the application of export controls for proliferation reasons. In calendar year 2009, the department reviewed 77 applications for otherwise non controlled items for missile "catch-all" related concerns. These applications were valued at 23.5 million dollars.

In addition, the EAR contains an Entity List that identifies specific end-users in countries throughout the world that pose a proliferation concern. Many of these end-users have been listed because of missile proliferation concerns. For most listed end-users, a license is required for all exports and reexports of items subject to the EAR.

The catch-all controls also go beyond control of items and extend to the activities of U.S. persons. Under the EAR, U.S. persons may not perform any contract, service, or employment that they know will directly assist in chemical and biological weapons or missile activities in or by certain countries. Our regulations also include prohibitions against exports of dual-use items, software or technology to sanctioned countries as designated by Congress.

Finally, our controls also target non-state actors. The EAR prohibits exports and reexports of any items to persons designated by the Department of the Treasury as Specially Designated Global Terrorists, Specially Designated Terrorists, or Foreign Terrorist Organizations. The Department of Commerce also maintains an extensive system of unilateral anti-terrorism controls, in addition to the controls imposed on items that are controlled for MT and other reasons. These controls are intended to keep even low-level goods and technologies out of the hands of the most dangerous actors.

It is also important to note our outreach program to U.S. industry. The government alone cannot protect our security interests in this globalized world. It is essential that the public and private sector combine their strengths to confront the threats to our economic and national security. The Department of Commerce has an extensive outreach program to inform U.S. industry of their export obligations and explain the scope of export controls to all exporters. Most U.S. companies are strongly committed to protecting our national security and they therefore seek to achieve excellent compliance with our laws. It is therefore imperative that those who could supply sensitive items to end-users of concern understand their obligations and the importance of compliance.

Thus, in addition to implementing our international commitments under the MTCR, the United States has in place a comprehensive program of additional measures to prevent the proliferation of missile systems capable of delivering WMD to countries of concern or terrorists.

C. Export Control Enforcement

BIS's Export Enforcement team, along with the Department of Homeland Security's Bureau of Immigration and Customs Enforcement, and the Federal Bureau of Investigation, enforce controls on dual-use exports. These agencies, through investigations of suspected violations of law and regulations, and the interdiction of suspected illicit shipments, have provided the necessary evidence to successfully prosecute both criminal and civil cases on export violations. Our multilateral controls also provide a strong framework for cooperative enforcement efforts overseas when such efforts call for an international approach.

The following highlights recent cases involving UAV's from the Department of Commerce's Office of Export Enforcement and the Department of Justice.

Aviation Services International

On September 24, 2009, Aviation Services International BV (ASI), an aircraft supply company in the Netherlands, Robert Kraaipoel, Director of ASI, Neils Kraaipoel, sales manager of ASI, and Delta Logistics pled guilty in U.S. District Court in Washington, DC to charges related to a conspiracy to illegally export aircraft components and other U.S.-origin commodities to entities in Iran, via the Netherlands, the United Arab Emirates and Cyprus. Between October 2005 and October 2007, the defendants received orders from customers in Iran for U.S.-origin items, including video recorder units for end use in Unmanned Aerial Vehicles, then contacted companies in the United States and negotiated purchases on behalf of the Iranian customers. The defendants provided false end-user certificates to U.S. companies to conceal the true end-users in Iran. The defendants caused U.S. companies to ship items to ASI in the Netherlands or other locations in the United Arab Emirates and Cyprus which were then repackaged and transshipped to Iran. In a related case, ASI, Robert Kraaipoel and Niels Kraaipoel settled administrative charges with BIS that included, in part, ASI and Robert Kraaipoel being placed on BIS's Denied Persons List for seven years. Niels Kraaipoel agreed to a three year denial of his export privileges that would be suspended pending no future export violations.

ARC International

On February 3, 2010, Harold Hanson (Hanson) and Nina Yaming Qi Hanson (Qi) were sentenced in U.S. District Court in the District of Columbia. Qi was sentenced to 105 days in jail with credit for time served, placed on one year of supervised release, ordered to pay a fine of $250 and a $100 special assessment fee and ordered to attend a U.S. Department of Commerce sponsored export education training program. Hanson was sentenced to 24 months probation, required to pay a fine of $250 and a $100 special assessment fee, ordered to perform 120 hours of community service, and also ordered to attend a U.S. Department of Commerce sponsored export training program. On November 13, 2009, Hanson and Qi pleaded guilty to making false statements. On March 12, 2009, a federal grand jury in the District of Columbia returned an indictment charging Qi, her husband Hanson (an employee at Walter Reed Army Medical Center), and a Maryland company, ARC International, LLC, with illegally exporting miniature Unmanned Aerial Vehicle (UAV) Autopilots controlled for national security reasons to a company in the People's Republic of China.

Mayrow General Trading

In September 2008, a federal grand jury in Miami, FL, returned a Superseding Indictment charging eight individuals and eight corporations in connection with their participation in conspiracies to export U.S.-manufactured commodities to prohibited entities and to Iran. They were charged with conspiracy, violations of the International Emergency Economic Powers Act and the United States Iran Embargo, and making false statements to federal agencies in connection with the export of thousands of U.S. goods to Iran. Charges against defendant Majid Seif, also known as Mark Ong, and Vast Solutions alleged that Seif and Vast exported radio control devices and accessories used in Unmanned Aerial Vehicles from a Singapore firm to Malaysia. The radio control devices were then shipped to Iran.

Landstar/Yi-Lan Chen

On February 3, 2010, Yi-Lan Chen, also known as Kevin Chen, was arrested on charges of illegally exporting commodities for Iran's missile program. According to the affidavit filed in support of the criminal complaint, Chen caused dual use goods to be exported from the U.S., including P200 Turbine Engines, which the investigation revealed were for end users in Iran. The P200 Turbine Engines are designed for use as model airplane engines but can also be used to operate Unmanned Aerial Vehicles and military target drones.

Conclusion

The Department of Commerce believes the issue of missile proliferation has never been as important to our national security interests as it is now. A comprehensive export control system is already in place to protect our national security. As noted above, the Department of Commerce is committed to enhancements to that system as needed to ensure it continues to protect our national security.

---

[1] We note that the MTCR is not the only international regime that works to prevent the spread of missile-related technologies. The Wassenaar Arrangement, the multilateral export control regime responsible for controls on conventional weapons and related items with both civilian and military (dual-use) applications, has recently imposed complementary controls on the export of UAVs.

FOR OFFICIAL USE ONLY
UNTIL RELEASED BY THE
HOUSE COMMITTEE ON
OVERSIGHT AND GOVERNMENT
REFORM

TESTIMONY OF

DYKE D. WEATHERINGTON

DEPUTY DIRECTOR, UNMANNED WARFARE

OFFICE OF THE UNDER SECRETARY OF DEFENSE

(ACQUISITION, TECHNOLOGY & LOGISTICS)

BEFORE THE UNITED STATES HOUSE

COMMITTEE ON

OVERSIGHT AND GOVERNMENT REFORM

SUBCOMMITTEE ON

NATIONAL SECURITY AND FOREIGN AFFAIRS

March 23, 2010

FOR OFFICIAL USE ONLY
UNTIL RELEASED BY THE
HOUSE COMMITTEE ON
OVERSIGHT AND GOVERNMENT
REFORM

Acquisition of Department of Defense Unmanned Aircraft Systems

Mr. Dyke D. Weatherington
Deputy Director, Unmanned Warfare
Office of the Under Secretary of Defense
(Acquisition, Technology and Logistics)

Chairman Tierney, Congressman Flake and Members of the Committee:

Thank you for the opportunity to appear before you today to discuss Department of Defense (DoD) unmanned aircraft system (UAS) acquisition programs, specifically, Department initiatives to achieve greater commonality and efficiencies. My testimony will address the full spectrum of DoD UAS, not just the larger (Groups 3-5) aircraft addressed by the Government Accountability Office (GAO) in their July 2009 report. The distinction is important because we have pursued opportunities for commonality and efficiency successfully across the full range of DoD unmanned aircraft, including small UAS. Table 1 is included to identify the broad diversity of DoD UAS supporting a wide range of warfighter needs. Classifications are based on aircraft weight, operating altitude, and speed. [1]

UAS Category	Max. Gross Takeoff Weight (lbs)	Normal Operating Altitude (ft)	Speed (KIAS[2])
Group 1	0-20	< 1200 AGL[3]	100
Group 2	21-55	< 3,500 AGL	<250
Group 3	<1320	< 18,000 MSL[4]	Any
Group 4	>1320		Any
Group 5		> 18,000 MSL	

Table 1: JUAS CONOPS UAS Categories[5]

---

[1] Specifics on the methodology employed by the JUAS COE to determine groupings are found in the Joint UAS Concept of Operations, 2nd Edition, November 2008.
[2] Knots (Nautical Miles per Hour) Indicated Airspeed
[3] Above Ground Level
[4] Mean Sea Level
[5] Lighter than air vehicles are classified by the highest of their operating attributes.

The GAO Report, "Defense Acquisitions: Opportunities Exist to Achieve Greater Commonality and Efficiencies among Unmanned Aircraft Systems," released in July 2009, reviewed DoD UAS Groups 3-5. The GAO had five recommendations. The Department partially concurred with the recommendation to conduct a rigorous and comprehensive analysis of requirements for current UAS and to develop a strategy for making systems and subsystems among those programs more common. At the time of the review, the UAS Task Force with support from the Joint Requirements Oversight Council had already completed a comprehensive analysis of the potential for commonality between Air Force Predator and Army Extended Range Multi Purpose UAS. Since the report was released, the UAS Task Force in coordination with the Joint Staff conducted a rigorous review of the Navy Broad Area Maritime Surveillance (BAMS) and Air Force Global Hawk programs to evaluate opportunities for achieving greater commonality and joint effectiveness. We have completed that analysis along with one addressing Signals Intelligence (SIGINT) payload commonality. I will address the findings of both these analyses later in my testimony. The Department concurred with remaining four recommendations which included requiring the Military Departments to identify and document in their acquisition plans and strategies specific areas where commonality can be achieved, to take an open systems approach to product development, to conduct a quantitative analysis that examines the costs and benefits of various levels of commonality, and to establish a collaborative approach and management framework to periodically assess and effectively manage commonality.

Since the GAO released its report, the Department has completed its 2010 Quadrennial Defense Review (QDR), and the President has submitted his Fiscal Year (FY) 2011 budget. The QDR highlights the warfighters' need for increased intelligence, surveillance, and reconnaissance (ISR) and force protection capabilities, and the budget reflects the Department's increased investment to meet that need. This investment is consistent with the Acquisition Reform goal in DoD's High Priority Performance Goals presented in the Analytic Perspectives volume of the President's FY 2011 Budget.

The Department's investment and operation in UAS continues to increase as the demand for the wide range of UAS capabilities expands. The DoD annual budget for development and procurement of UAS has increased from $1.7 billion in FY 2006 to over $4.2 billion in FY 2010. During the same time period DoD UAS operations have grown from 165,000 hours to over 550, 000 hours annually (as shown in Figure 1 on page 4), and the unmanned aircraft inventory has increased from less than 3,000 to over 6,500.

The Department is making significant investments in UAS, and that is projected to grow significantly over the next five years. Achieving commonality, interoperability, and joint efficiencies in development, production, and operations and support is critical to controlling cost and delivering interoperable and reliable systems to the warfighters with the capabilities they need to win. We continue to improve the Defense Acquisition System, and have formed a UAS Task Force to jointly address critical UAS technical and acquisition issues to enhance operations and enable interdependencies, commonality and other efficiencies.

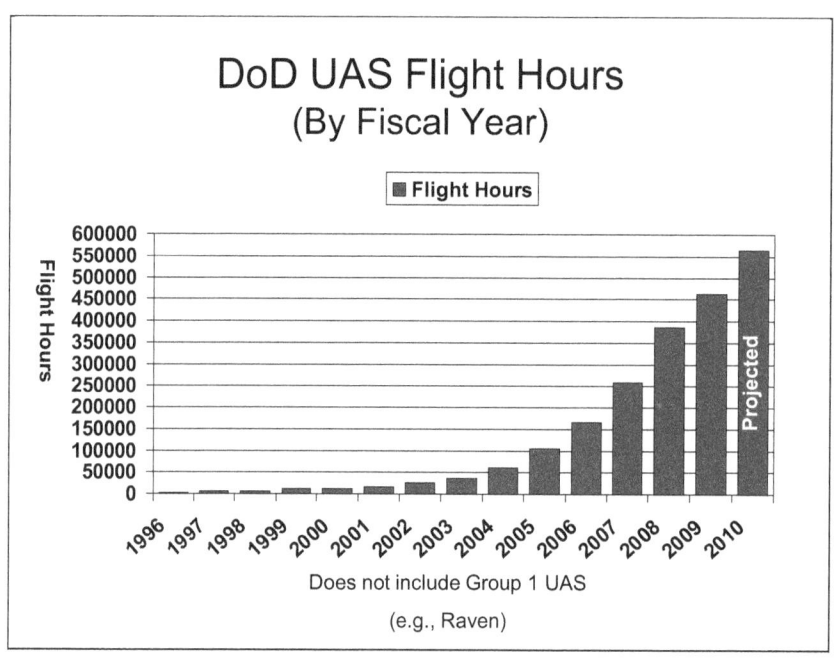

Figure 1 DoD UAS Flight Hours

I will now update you on the DoD UAS acquisition and summarize our efforts to increase commonality and achieve additional efficiencies.

Update on DoD UAS Programs

Overview of UAS Programs

In FY 2010, the Department made a commitment to grow Air Force Predator/Reaper combat air patrols (CAPs) to 50 by FY 2011. The Air Force is on track to achieve this goal and will continue to expand force structure to support up to 65 CAPs by FY 2013. The Army is expanding all classes of UASs, including the accelerated production of the Predator-class Extended Range Multi-Purpose (ER/MP) UAS and upgrading its Shadow UAS. The Army is embedding UAS in all its Brigade Combat Teams. In addition to the Quick Reaction Capability of four ER/MP aircraft already fielded in Iraq, the Army will field a second Quick Reaction Capability to Afghanistan in FY10. The Army also plans to field 13 ER/MP systems of 12 aircraft each to each Combat Aviation Brigade, starting in FY 2011. The Navy is in the Engineering and Manufacturing Development phase for its Broad Area Maritime Surveillance (BAMS) UAS, and is introducing sea-based UASs with its Vertical Take Off Unmanned Aerial Vehicle (VTUAV) and its Small Tactical Unmanned Aircraft System (STUAS) programs. The Navy plans to award the STUAS contract during the next quarter. All Military Departments and the U.S. Special Operations Command (SOCOM) are operating the hand-launched Raven with over 4,700 aircraft delivered to the warfighter.

<u>Overview of Department Initiatives to Increase Commonality and Efficiencies</u>

The *Department of Defense Report to Congress on Common Control Stations and Payloads for Manned and Unmanned Aircraft Systems (UAS)* was forwarded to the congressional defense committees on June 25, 2009, in response to Section 144 of the National Defense Authorization Act for FY 2009. It describes the Department's initiatives to increase commonality and efficiencies for manned and unmanned aircraft systems. The Department of Defense Instruction (DoDI) 5000.02, dated December 8, 2008, improves the Defense Acquisition System by providing increased management focus early in program creation; this enables the Department to identify opportunities for commonality and efficiencies across Military Departments and programs at the Material Development Decision milestone in the acquisition life cycle. The UAS Task Force, led by the Under Secretary of Defense for Acquisition, Technology and Logistics (USD(AT&L)) with members from the Military Departments, the Office of the Secretary of Defense (OSD), the Joint Staff, and Joint Forces Command, continues to coordinate critical UAS issues to enhance operations, enable interdependencies, and facilitate delivery of UAS capability to the warfighter. Key areas of UAS TF activity include civil airspace integration, frequency spectrum and bandwidth utilization, information assurance and encryption, and interoperability among ground stations and airframes across Military Departments with payload and sensor management transparencies.

In addition to working within the Department toward improving commonality and efficiencies among UAS, the DoD has joined with the Department of Transportation, the Department of Homeland Security, and the National Aeronautics and Space Administration to form a UAS Executive Committee (ExCom) to rapidly improve public UAS integration into the National Airspace System (NAS). DoD UAS require NAS access for training, development, and operations. Access is critical to supporting multiple warfighters with the ISR and force protection capabilities required to fight and win our current overseas contingency operations. We continue to engage directly with the Federal Aviation Administration (FAA) via the UAS ExCom to identify and resolve issues surrounding integration of UAS into the NAS. The FAA hosted the first UAS ExCom meeting in October 2009.

I will now summarize the Department's progress toward achieving greater commonality and efficiencies for the Group 3-5 UAS programs reviewed for the GAO Report, and also Group 1-2 UAS programs that were not part of the study.

<u>RQ-4 Global Hawk / Broad Area Maritime Surveillance (BAMS) Joint Efficiencies</u>

The Air Force Global Hawk and Navy BAMS UAS programs are just two of our unmanned systems that continue to pursue opportunities for greater commonality and efficiency. The Department understands that there exists a significant cost and operational benefit in leveraging system commonality where appropriate in the acquisition of our weapon systems. Specifically, within Global Hawk and BAMS UAS programs we are stressing several developmental areas to increase commonality both in weapon system design and during military operations. We are utilizing common production facilities at Palmdale, California, for both Global Hawk and BAMS EMD aircraft. Additionally, we are addressing critical flight safety issues of a new and technologically sound approach to operate UAS in the national airspace through the joint development and acquire a sense and avoid system. The Air Force and the Navy have agreed to jointly acquire a common radar as part of that system and the Military Departments are cooperating in expanding our aircraft certification process to include unique UAS elements. As the Global Hawk and BAMS systems share common elements and both operate beyond-line-of-sight long endurance missions, we are fully leveraging these similarities by re-architecturing the mission ground stations to streamline operations and mission effectiveness. The payoff for this production decision of ground station commonality is that we expect to enact total cost efficiencies for the equipment, sustainment, and training as well as reap operational efficiencies for the warfighter. OSD and the Joint Staff are actively coupling these acquisition efforts; together we monitor and encourage all efforts to increase joint efficiencies. Our efforts to increase commonality are documented in the Acquisition Decision Memorandums and Joint

Requirements Oversight Council Memorandums that direct these two Military Departments to work together to gain even more elements of commonality, such as incorporating a similar anti-ice technology for both Global Hawk and BAMS UAS aircraft to expand operations in icing conditions. Other specific directed actions include efforts to increase joint efficiencies in the areas of operations and support. Examples include: common basing; consolidated maintenance; beyond-line-of-sight command and control; processing, exploitation and dissemination functions; training; and supply chain management. In summary, Global Hawk and BAMS UAS are two unmanned systems that embrace the Department's efforts to use commonality and joint efficiency efforts to lessen total program costs, leverage benefits derived from joint technology development, and implement common systems when favorable.

Electro Optical Infrared (EO/IR) Sensor Payload for Predator and ER/MP

The Air Force and the Army have converged on a common electro optical infrared (EO/IR) sensor payload for Predator and ER/MP. The sensor will provide high definition (HD) full motion video and enhanced target location accuracy (TLA) supported by a fully digital infrastructure. The Military Departments plan to procure from a common contract with the initial orders for the HD EO upgrade placed the third quarter of this fiscal year, and deliveries to the Army in FY 2011. HD IR and TLA will be added sequentially and the fully upgraded sensor will be fielded in FY 2014 by both Military Departments.

Signals Intelligence (SIGINT) Payload

The Department, recognizing that requirements for the Air Force's Airborne Signals Intelligence Payload (ASIP) and the Army's Tactical Signals Intelligence

Payload (TSP) sensors appear similar, determined a UAS SIGINT sensor business case analysis was necessary to provide an independent assessment of the cost, schedule, and performance implications of migrating to a highly common SIGINT system for the Predator, ER/MP, and Reaper fleets. The UAS SIGINT business case analysis includes a comparison of the cost versus capability for the ASIP and the TSP sensors to meet the broad range of Military Departments', and Component and Combatant Commanders' SIGINT capability requirements, and a technical assessment of each system providing context for the requirements and desired performance parameters, industrial base issues and program oversight. Special consideration was given towards technology maturity, integration, and employment criteria to better understand the operational risks and future challenges. The Department expects to finalize its assessment of the findings next month, and will include its assessment in its *Report to Congress on Unmanned Aircraft Systems (UAS)-based Signals Intelligence (SIGINT) Payload.*

Shadow UAS

The Army and the United States Marine Corps (USMC) continue to procure and operate the common Shadow UAS while incorporating program improvements. The Army Procurement Objective has increased from the initial 41 in the original December 26, 2002 Acquisition Program Baseline to 102 in July 2008. All of the aircraft are Shadow 7B variants with several different configurations. Program has been in full rate production since 2002 with current Army production ending in FY 2011. The USMC also operates the Shadow system and is procuring 13 systems utilizing the Army production contract. The USMC utilizes the Army training base and sustainment capability to reduce cost and increase efficiency.

The Shadow program has undergone numerous upgrades since the full rate production decision in FY 2002. Major upgrades completing test include incorporation of a laser designator (LD) in the EO/IR sensor, engine electronic fuel injection (EFI), and aircraft re-wing. LD provides precision targeting capability for laser munitions including Hellfire equipped Apaches, and other aircraft capable of carrying laser seeking munitions. EFI improves engine performance, fuel consumption, and reliability, and eliminates carburetor icing. Re-wing improves endurance from five to eight hours and enables future incorporation of Tactical Common Data-Link. All these Army funded upgrades will migrate into both Army and USMC Shadow force structure.

Groups 1-2 UAS Programs

While not included in the GAO's recent review, I would like to take this opportunity to tell a success story with respect to small UAS (SUAS) commonality. Immediately following September 11, 2001, the Military Departments and SOCOM procured a number of different small hand-launched UAS. The SUAS proved to be a low cost, highly effective force protection system, and many different types were procured; Pointer, Dragon Eye, Swift, Raven A, and Desert Hawk were the primary systems fielded, but there were others as well. Today, all the Military Departments and SOCOM are procuring the Raven B system using the same contract and realizing cost savings while gaining improved performance. Today's Raven B has greater capability than the original system; improvements include greater aircraft endurance, and improved sensor, communications and ground station capabilities. There are over 1,500 Raven B systems fielded; each system includes 3 aircraft.

Conclusion

In closing Mr. Chairman, the Department's investment in UAS is projected to continue to grow. We recognize that achieving commonality, interoperability, and joint efficiencies in development, production, and operations and support is critical to controlling cost and delivering interoperable and reliable systems to the warfighters with the capabilities they need to win. We continue to improve the Defense Acquisition System, and have formed a UAS Task Force to jointly address critical UAS issues to enhance operations and enable interdependencies, commonality and other efficiencies. These Task Force efforts are consistent with the Acquisition Reform goal in DoD's High Priority Performance Goals presented in the Analytic Perspectives volume of the President's FY 2011 Budget.

Thank you for the opportunity to testify before the Committee. I would be happy to answer any questions you and the Members of the Committee may have.

United States Government Accountability Office

GAO

Testimony
Before the Subcommittee on National Security and Foreign Affairs, Committee on Oversight and Government Reform, House of Representatives

For Release on Delivery
Expected at 2:00 p.m. EST
Tuesday, March 23, 2010

DEFENSE ACQUISITIONS

DOD Could Achieve Greater Commonality and Efficiencies among Its Unmanned Aircraft Systems

Statement of Michael J. Sullivan, Director
Acquisition and Sourcing Management

GAO-10-508T

March 23, 2010

DEFENSE ACQUISITIONS

DOD Could Achieve Greater Commonality and Efficiencies among Its Unmanned Aircraft Systems

Highlights of GAO-10-508T, a testimony before the Subcommittee on National Security and Foreign Affairs, Committee on Oversight and Government Reform, House of Representatives

Why GAO Did This Study

For the last several years, the Department of Defense (DOD) has planned to invest billions of dollars in development and procurement of unmanned aircraft systems. In its fiscal year 2011 budget request the department indicated a significant increase in these investments, expecting to need more than $24 billion from 2010 through 2015. DOD recognizes that to leverage its resources more effectively, it must achieve greater commonality among the military services' unmanned aircraft system acquisition programs.

This testimony is based primarily on GAO's July 2009 report (GAO-09-520) which examined 10 unmanned aircraft acquisition programs: eight unmanned aircraft systems—Global Hawk, Reaper, Shadow, Predator, Sky Warrior, Fire Scout, Broad Area Maritime Surveillance, and Unmanned Combat Aircraft System-Demonstration; and two payload development programs—Multi-Platform Radar Technology Insertion Program, and Airborne Signals Intelligence Payload. The testimony focuses on: 1) the cost, schedule, and performance progress of the 10 programs as of July 2009; 2) the extent to which the military services collaborated and identified commonality among the programs; 3) factors influencing the effectiveness of the collaboration; and, 4) recent DOD investment decisions related to these acquisitions.

View GAO-10-508T or key components.
For more information, contact Michael Sullivan at (202)512-4841 or sullivanm@gao.gov.

What GAO Found

Most of the 10 programs reviewed had experienced cost increases, schedule delays, performance shortfalls, or some combination of these problems. The programs' development cost estimates increased by more than $3 billion collectively, or 37 percent, from initial estimates. Procurement funding requirements for most programs also increased, primarily because of increases in numbers of aircraft being procured, changes in system requirements, and upgrades and retrofits to fielded systems. Procurement unit costs increased by an average of 12 percent, with three aircraft programs experiencing unit cost increases of 25 percent or more. Four programs reported delays of 1 year or more in delivering capability to the warfighter. Global Hawk, Predator, Reaper, and Shadow had been used in combat operations with success and lessons learned, but had been rushed into service in some cases, leading to performance issues and delays in development and operational testing and verification.

Programs collaborated and identified areas of commonality to varying degrees. The Marine Corps was able to avoid the cost of initial system development and quickly deliver useful capability to the warfighter by choosing to procure existing Army Shadow systems. The Navy expected to save time and money on Broad Area Maritime Surveillance (BAMS) by using Air Force's Global Hawk airframe, and payloads and subsystems from other programs. However, Army and Air Force had not collaborated on their Sky Warrior and Predator programs, and might have achieved greater savings if they had, given that Sky Warrior is a variant of Predator and being developed by the same contractor. DOD encouraged more commonality between these programs.

Although several programs achieved airframe commonality, service-driven acquisition processes and ineffective collaboration were key factors that inhibited commonality among subsystems, payloads, and ground control stations, raising concerns about potential inefficiencies and duplication. Despite DOD's efforts to emphasize a joint approach to identifying needs and commonality among systems, most of the programs assessed continued to pursue service-unique requirements. The services also made independent resource allocation decisions to support their unique requirements. DOD had not quantified the costs and benefits associated with pursuing commonality among these programs, and efforts to collaborate had produced mixed results. However, in order to maximize acquisition resources and meet increased demand, Congress and DOD have continued to push for more commonality.

Since July 2009, DOD has made several investment decisions regarding unmanned aircraft systems, which in general, reflect increased emphasis on developing advanced capabilities and acquiring larger numbers of specific systems. However, the decisions do not appear to focus on increasing collaboration or commonality among the programs.

United States Government Accountability Office

Mr. Chairman and Members of the Subcommittee:

Thank you for this opportunity to discuss GAO's recently issued report on the Department of Defense's (DOD) unmanned aircraft systems (UAS) acquisition efforts.[1] From 2002 through 2008, the number of unmanned aircraft in the DOD's inventory increased from 167 to more than 6,000 as a result of the department's efforts to meet the growing demand from the warfighters for these capabilities. DOD has noted that meeting this demand has been difficult because of the dynamic nature of supporting ongoing combat operations in Iraq and Afghanistan, while at the same time developing new and emerging capabilities. At the time of our report in July 2009, the department was planning to invest more than $16 billion from 2008 through 2013 to develop and procure additional unmanned aircraft systems. More recently, the fiscal year 2011 defense budget request indicates a significant increase in DOD's unmanned aircraft investment plans. However, the growing number of national priorities competing for federal dollars will continue to challenge DOD's efforts to meet escalating demands for unmanned systems.

DOD recognizes that to more effectively leverage its acquisition resources, it must achieve greater commonality and efficiency among the military services' various unmanned system acquisition programs. In fact, DOD states in its Unmanned Systems Roadmap, that there is the potential for an unprecedented level of collaboration to meet capability needs and reduce acquisition costs by requiring greater commonality among the military services' unmanned systems. Although achieving commonality can be difficult, we have reported in the past that taking an open systems[2] approach and designing systems with common subsystems and components can reduce both production and life cycle costs as well as improve interoperability among systems. For maximum benefit, commonality should be incorporated into the design of a system when requirements are being established. Unmanned aircraft systems can potentially achieve commonality in design and development, ranging from a complete system, including the ground control segment, to a subsystem

---

[1]GAO, *Defense Acquisitions: Opportunities Exist to Achieve Greater Commonality and Efficiencies among Unmanned Aircraft Systems*, GAO-09-520 (Washington, D.C.: July 30, 2009).

[2]Open systems allow the use of commercially available and widely accepted standard products from multiple vendors, rather than developing unique components.

or component, as well as commonality in production facilities, tooling, and personnel.

My statement today focuses on (1) the cost, schedule, and performance progress of selected unmanned aircraft acquisition programs as of July 2009; (2) the extent to which the military services had collaborated and identified commonality among those programs; (3) the key factors influencing the effectiveness of their collaboration; and (4) recent DOD investment decisions related to unmanned aircraft acquisitions. It is primarily drawn from our July 2009 report that examined 10 acquisition programs: eight unmanned aircraft programs and two payload programs. We conducted this performance audit from August 2008 to July 2009 in accordance with generally accepted government auditing standards. Those standards require that we plan and perform the audit to obtain sufficient, appropriate evidence to provide a reasonable basis for our findings and conclusions based on our audit objectives. We believe that the evidence obtained provides a reasonable basis for our findings and conclusions based on our audit objectives.

Summary

Once fielded, unmanned aircraft have proven quite valuable to the warfighter. On the other hand, most of the unmanned aircraft programs we reviewed had experienced cost increases, schedule delays, performance shortfalls, or some combination of these problems. Development cost estimates for the 10 programs we assessed had collectively increased more than $3 billion (37 percent in 2009 dollars) from initial estimates. Procurement funding requirements had also increased for most programs, primarily because of increases in the number of aircraft being procured, changes in system requirements, and upgrades and retrofits to equip fielded systems with capabilities that had been deferred. Overall, procurement unit costs increased by 12 percent, with three aircraft programs experiencing unit cost increases of 25 percent or more. Four programs had reported delays of 1 year or more in delivering capability to the warfighter. While the Global Hawk, Predator, Reaper, and Shadow systems had been used in combat operations with notable success and key lessons learned, they had been rushed into service in some cases, leading to performance issues and delays in development and operational testing and verification.

We found varying degrees of collaboration and commonality among DOD's unmanned aircraft acquisition programs. The Marine Corps was able to avoid the cost of initial system development and quickly deliver useful capability to the warfighter by choosing to procure existing Army Shadow systems. The Army and Navy had settled on many common requirements between their Fire Scout systems, which had the potential to gain them efficiencies. However, in January 2010 the Army notified the Congress that it had terminated its Fire Scout program because the aircraft was no longer required. In another case, the Navy expected to save time and money on its Broad Area Maritime Surveillance (BAMS) system by using the existing Air Force Global Hawk airframe, with payloads and subsystems from various other programs. In contrast, the Army and Air Force had not effectively collaborated on their Sky Warrior and Predator programs, and greater commonality could have been achieved given that the Sky Warrior is a variant of the Predator and is being developed by the same contractor. At the time of our review, DOD officials continued to press for more commonality between these two programs.

Service-centric requirements and funding, and ineffective collaboration were key factors that resulted in the limited achievement of commonality. While several unmanned aircraft programs had achieved airframe commonality, most were pursuing service unique subsystems, payloads, and ground control stations. Despite DOD's efforts to emphasize a joint approach to identifying and prioritizing warfighting needs and to encourage commonality among programs, the services continued to establish service-unique requirements—some of which have raised concerns about possible inefficiencies caused by unnecessary duplication. Likewise, DOD's funding process gives the individual services the responsibility and authority to independently make resource allocation decisions to support their respective requirements. At the time of our review, DOD officials had not quantified the associated costs or benefits of pursuing increased commonality among unmanned aircraft programs, and service efforts to collaborate had produced mixed results. However, Congress and DOD continued to push for more commonality, which could maximize acquisition resources and help meet increased demand.

Since July 2009, when our report was issued, DOD has made several key investment decisions regarding unmanned aircraft systems that are contained in the 2010 Quadrennial Defense Review, DOD's fiscal year 2011 budget request, and DOD's Aircraft Investment Plan (2011-2040). In general, these decisions reflect increased emphasis on developing more advanced unmanned aircraft capabilities and acquiring larger numbers of specific systems. However, they do not appear to focus on increasing collaboration or commonality among unmanned aircraft programs.

Background

Unmanned aircraft systems generally consist of (1) multiple aircraft, which can be expendable or recoverable and can carry lethal or non-lethal payloads; (2) a flight control station; (3) information and retrieval or processing stations; and (4) in some cases, wheeled land vehicles that carry launch and recovery platforms. DOD categorizes these systems based on key characteristics including weight and operating altitude. While there were many small, less expensive unmanned aircraft in DOD's portfolio, our review focused on the larger, more costly programs. At that time, these programs accounted for more than 80 percent of DOD's total investment in unmanned aircraft from fiscal year 2008 through fiscal year 2013.[3] DOD's 2011 budget request indicates that the department plans to invest nearly $25 billion from 2010 through 2015 in development and procurement of the unmanned aircraft systems we reviewed. Table 1 details many of the key characteristics and funding requirements of those systems. See appendix I for additional program data.

Table 1: Characteristics and Funding Requirements of Selected Unmanned Aircraft Systems

(Then year dollars in millions)

Aircraft	Gross Weight (pounds)	Maximum Altitude (feet)	Imagery Intelligence	Signals Intelligence	Weapons	Total Investment Funding (FY10-FY15)
Reaper	10,500	50,000	X	X	X	$8,354.7
Global Hawk[a]	32,250	60,000	X	X		5,130.1
BAMS	32,250	60,000	X			3,783.9
Sky Warrior	3,200	25,000	X		X	3,306.1
Shadow	375	15,000	X			1,781.4
UCAS-D	46,000	40,000	n/a	n/a	n/a	1,056.4
Predator	2,250	25,000	X	X	X	829.5
Fire Scout[b]	3,150	20,000	X	X		472.4

Sources: DOD, Unmanned Systems Roadmap 2007 – 2032 and BAMS Program Office

[a]Global Hawk characteristics presented in this table refer to the RQ-4B.

[b]Fire Scout data presented here are for the Navy program only.

Note: While we also assessed the Navy's Unmanned Combat Aircraft System Demonstration (UCAS-D) as part of our review, UCAS-D is a demonstration effort and will not be equipped with any mission payloads.

---

[3]The programs we focused on are often referred to as tactical-level and theater-level systems.

Unmanned Aircraft Acquisitions Have Experienced Cost Growth, Schedule Delays, and Performance Problems

Despite the proven success of unmanned aircraft on the battlefield and the growing demand for the aircraft, these acquisitions continued to incur cost and schedule growth. The cumulative development cost for the 10 programs we reviewed increased by over $3 billion, or 37 percent, from initial estimates. While 3 of the 10 programs had little or no development cost growth and one had a cost reduction, six experienced substantial growth ranging from 60 to 264 percent. This cost growth was in large part the result of changes in program requirements and system designs after initiating development. Many of the programs began system development with unclear or poorly defined requirements, immature technologies, and unstable designs—problems we have frequently found in other major acquisition programs.[4] For example, in 2001, the Air Force began the Global Hawk program based on knowledge gained from a demonstration program, and planned to incrementally integrate more advanced technologies over time. Within a year, however, the Air Force fundamentally restructured and accelerated the program to pursue a larger, unproven airframe with a multimission capability that relied on immature technologies. The final design of the new airframe required more substantial changes than expected. These changes ultimately drove development costs up nearly threefold.

Procurement costs also increased for 6 of the 7 systems that reported procurement cost data. Although in large part the cost increases were due to the planned procurement of additional aircraft, many programs had also experienced unit cost increases independent of quantity. As detailed in table 2, overall procurement unit costs increased by 12 percent on average, with three programs experiencing unit cost growth of 25 percent of more. The Reaper and Shadow had unit cost growth despite increased quantities. Reaper's unit costs increased in part because requirements for missiles and a digital electronic engine control were added—resulting in design changes and increased production costs. Unit cost increases in the Shadow program were largely the result of upgrades to the airframe that were needed to accommodate the size, weight, and power requirements for integrating a congressionally mandated data link onto the aircraft.[5] Furthermore, the Army is retrofitting fielded systems with capabilities that it had initially deferred, such as a heavy fuel engine.

---

[4]GAO, *Defense Acquisitions: Assessments of Selected Weapon Programs,* GAO-09-326SP (Washington, D.C.: Mar. 30, 2009).

[5]National Defense Authorization Act for Fiscal Year 2006, Pub. L. No. 109-163, § 141.

Table 2: Cost and Quantity for Selected Unmanned Aircraft Systems (as of July 2009)

2009 dollars in millions

Aircraft	Estimated development cost	Initial procurement cost estimate	Initial quantity	Current procurement cost estimate	Current quantity	Percent procurement unit cost change
Global Hawk	$3,657.5	$4,171.4	63	$5,929.7	54	66
Reaper[a]	385.5	508.7	33	2,405.7	118	32
Shadow	356.6	447.0	160	1,640.7	460	28
Fire Scout[b]	605.0	1,625.1	168	1,743.0	168	7
BAMS	3,049.1	9,048.6	65	9,048.6	65	0
Sky Warrior	568.5	647.5	48	1,614.2	132	-9
Predator	428.2	642.8	48	2,546.4	320	-41
UCAS-D[c]	1,474.9	n/a	n/a	n/a	n/a	n/a
Total	$10,525.3	$17,091.1	585	$24,928.3	1,317	12 (average)

Sources: DOD (data); GAO (analysis and presentation)

[a]Initial procurement cost estimate provided for Reaper was based on 33 aircraft. However, the Air Force initially planned for 63 aircraft.

[b]Fire Scout data presented here are for the Navy program only.

[c]UCAS-D is a demonstration effort only, so the Navy was not projecting procurement funding or quantities.

A number of programs had experienced problems in both testing and performance, requiring additional development that contributed to the cost growth noted above. Four programs had experienced delays of 1 to nearly 4 years in achieving initial operational capability. Some of these delays resulted from expediting limited capability to the warfighter, while others were the result of system development and testing problems. For example, early demonstration and production Global Hawks were rushed into operational service. Program officials noted that as a result, the availability of test resources and time for testing were limited, which delayed the operational assessment of the original aircraft model by 3 years. Similarly, in February 2009, the Air Force reported that initial operational testing for the larger, more capable Global Hawk aircraft and the program's production readiness review had schedule breaches. Air Force officials cite the high level of concurrency between development, production, and testing; poor contractor performance; developmental and technical problems; system failures; and bad weather as key reasons for the most recent schedule breach.

Efforts to Collaborate and Identify Commonality Were Successful in Some Cases, While Not in Others

Consistent with DOD's framework for acquiring unmanned systems, some of the tactical and theater-level unmanned aircraft acquisition programs we reviewed had identified areas of commonality to leverage resources and gain efficiencies. For example, the Army and Marine Corps achieved full commonality in the Shadow program. In assessing options for replacing an aging tactical unmanned aircraft system,[6] the Marine Corps determined that the Army's Shadow system could meet its requirements for reconnaissance, surveillance, and target acquisition capabilities without any service-unique modifications. An official from DOD's Office of Unmanned Warfare emphasized that the Marine Corps believed that Shadow represented a "100 percent" solution. The Marine Corps also found that it could use the Army's ground control station to pilot the Shadow aircraft as well as other Marine Corps unmanned aircraft. A memorandum of agreement was established in July 2007 to articulate how the Marine Corps and the Army would coordinate to acquire Shadow systems.

By forgoing any service-unique modifications in order to achieve a high level of commonality, the Marine Corps avoided the costs of developing the Shadow. Additionally, the Marine Corps and Army are likely to realize some benefits in supporting and maintaining the systems because the components are interchangeable. The Army's Shadow program office agreed that commonality has allowed the two services to realize economies of scale while meeting each service's needs. According to an official at the Navy, the Marine Corps has been able to realize savings or cost avoidance in other areas such as administration, contracting, and testing, although quantitative data on these savings were not available.

In some cases, the services had collaborated to identify common configuration, performance, and support requirements, but ultimately were not maximizing efficiencies. For example, the Army and Navy had different data link requirements for their respective variants of Fire Scout, primarily because of the Army's requirement for its variant to operate within the Future Combat Systems network. According to the Fire Scout contractor, the Army's system could have been equipped with the same data link as the Navy Fire Scout, as well as the Army's Shadow and Sky Warrior systems, and placed into service sooner. Though the services had

---

[6]Shadow was identified as a replacement system for the Marine Corps Pioneer unmanned aircraft. Specifically, the cost for maintaining the Pioneer fleet was cited as a reason for selecting the Shadow system. The Marine Corps is considering a future replacement to the Shadow, which is not expected before 2015.

not agreed on a common data link, the Army and Navy had settled on common Fire Scout requirements for the air vehicle, engine, radar, navigation, and some core avionics subsystems requirements. The services had also agreed to use one contract to procure the airframe. However, in an information letter sent to members of Congress on January 11, 2010, the Army noted that it had terminated the Fire Scout portion of its FCS contract—following a decision by the Office of the Secretary of Defense (OSD) to cancel the FCS program—because analysis indicated that an improved Shadow system could meet future Army requirements, and the Fire Scout was no longer needed. Cancellation of the Army Fire Scout could lead to increased unit cost for the Navy variant.

Although the Navy BAMS and Air Force Global Hawk programs had identified commonalities between their airframes, the two programs had established different payload, subsystem, and ground station requirements. The Navy anticipated spending more than $3 billion to modify the Global Hawk airframe and ground stations, and to integrate Navy-specific payloads, including the radar. In addition, we found that the Navy had an opportunity to achieve greater efficiency in BAMS production. While production of the first two BAMS aircraft was planned to occur at the same California facility that produces Global Hawk, the remaining aircraft were expected to be produced at a facility in Florida. We pointed out that this approach might create duplication in production by staffing and equipping two facilities to conduct essentially the same work. At the time of our review the Navy had not assessed the costs or benefits of establishing a second production facility, and according to contractor officials, the official business case analysis would not be conducted for several years. Therefore, it was unclear whether any benefits of a second production facility would outweigh costs, such as additional tooling and personnel.

In contrast to the examples of the Shadow, Fire Scout, and BAMS / Global Hawk programs above, the Army and Air Force missed opportunities to achieve commonality and efficiencies between their Sky Warrior and Predator programs. In 2001, the Army began defining requirements for a replacement to the aging Hunter unmanned aircraft system, and decided to pursue the development of Sky Warrior. Both the Air Force and the Joint Staff responsible for reviewing Sky Warrior's requirements and acquisition documentation raised concerns about duplicating existing capability—specifically, capability provided by Predator. Nevertheless, the Army program received approval to forgo an analysis of alternatives that could have determined whether or not existing capabilities met its requirements. The Army noted that such an analysis was not needed and not worth the cost and effort. Instead, it conducted a source selection competition and began the Sky Warrior development program in 2005, citing battlefield commanders' urgent need for the capability. The development contract was awarded to the same contractor working with the Air Force to develop and produce Predators and Reapers. Since the Sky Warrior is a variant of the Predator, the two aircraft are assembled in the same production facility. Despite the establishment of a memorandum of understanding in 2006, direction from the Deputy Secretary of Defense in 2007 to combine their programs, and a subsequent memorandum of agreement, the Army and Air Force maintained separate programs and at the time of our review, had achieved little commonality.

Service-Centric Acquisition Processes and Ineffective Collaboration Have Reduced Opportunities for Commonality

While several of the unmanned aircraft programs we examined had achieved commonality at the airframe level, service-centric acquisition processes and ineffective collaboration resulted in service-unique subsystems, payloads, and ground control stations. Despite DOD's efforts to encourage a joint approach to identifying and prioritizing warfighting needs and to emphasize the need for commonality among the programs, we noted that the individual services continued to drive requirements and make independent resource allocation decisions. In many cases, the services had established requirements so specific that they demanded service-unique solutions, thereby precluding opportunities for commonality. Within DOD's funding system, each service has the responsibility and authority to prioritize its own budget, allowing it to make independent funding decisions to support unique requirements. Therefore, once a service concludes that a unique solution is warranted, the service has the authority to budget for that unique solution, to the exclusion of other solutions that might achieve greater commonality and efficiencies. While we recognized that service-unique requirements appeared to be necessary in some cases, one OSD official we spoke with emphasized concerns that some of the services' distinctions in requirements could lead to duplication and inefficiencies. However, OSD had not quantified the potential costs or benefits of pursuing various alternatives, including commonality.

In 2007, OSD established the Unmanned Aircraft Systems Task Force and the Office of Unmanned Warfare primarily to facilitate collaboration and encourage greater commonality among unmanned aircraft programs. While the two groups act as advisors and have implemented OSD's recommendations regarding areas where further commonality might be achieved key officials from these groups emphasized to us that they do not have direct decision-making or resource allocation authority.[7] OSD repeatedly directed the Army and Air Force to collaborate on their Sky Warrior and Predator programs, but the services continued to pursue unique systems. In response to OSD direction to merge their unique signals intelligence payload efforts into a single acquisition program, the Army and Air Force concluded that continuing their separate programs was warranted, and recommended that OSD direct an objective, independent organization—such as a federally funded research and development center—to conduct a business case analysis to assess the impact of merging the two programs.[8] Table 3 summarizes OSD's directions and the services' responses over the past few years.

---

[7] GAO recently reported (GAO-09-175) that the Under Secretary of Defense for Acquisition, Technology, and Logistics created the task force in 2007 to lead a DOD-wide effort to coordinate critical unmanned aircraft systems issues and develop a way ahead to enhance operations and streamline acquisitions.

[8] In a March 2010 meeting with the Air Force Predator and Reaper program office, program officials noted that the Air Force and Army are now pursuing a common sensor payload for their respective aircraft.

Table 3: OSD and Service Efforts to Achieve Predator and Sky Warrior Commonality

	OSD	Services
November 2006	Under Secretary of Defense for Acquisition, Technology and Logistics (AT&L) establishes goal for the programs to have a common aircraft, propulsion system, and avionics configuration	
September 2007	Deputy Secretary of Defense directs the services to combine the programs into a single acquisition program and to migrate to a single contract by October 2008	
February 2008		Army and Air Force program executive officers sign a memorandum of agreement
May 2008	Under Secretary of Defense for AT&L reiterates the Deputy Secretary of Defense's directive to combine the programs into a single acquisition program, states that fiscal year 2009 funds can only be used to purchase a common airframe, and expresses dissatisfaction with the progress made on achieving a common electro-optical and infrared sensor	
October 2008	Undersecretary of Defense for AT&L grants a waiver to the Air Force to buy 20 additional Predators, but also directs the Air Force to buy five common airframes and noted that no additional waivers would be granted	
January 2009	Deputy Under Secretary for Acquisition and Technology and the Deputy Under Secretary of Defense (Intelligence) for Portfolio, Programs, and Resources direct the services to conduct a comprehensive business case analysis to assess the impacts of migrating to a single signals intelligence payload acquisition program	
February 2009		Acting Assistant Secretary of the Army (Acquisition, Logistics, and Technology) and Assistant Secretary of the Air Force (Acquisition) issue a joint memorandum, noting that despite more than 15 months of work and a dozen meetings, neither service supports the assertion that a joint program makes sense, and recommend that an objective, independent agency or organization do the business case analysis

Source: GAO.

Congress and OSD took additional action in 2009 aimed at increasing collaboration and commonality among unmanned aircraft programs. In section 144 of the Duncan Hunter National Defense Authorization Act for Fiscal Year 2009, Congress directed "[t]he Secretary of Defense, in consultation with the Chairman of the Joint Chiefs of Staff, [to] establish a policy and an acquisition strategy for intelligence, surveillance, and reconnaissance payloads and ground stations for manned and unmanned aerial vehicle systems. The policy and acquisition strategy shall be applicable throughout the Department of Defense and shall achieve integrated research, development, test, and evaluation, and procurement commonality."[9] In an acquisition decision memorandum issued on February 11, 2009, the Under Secretary of Defense for Acquisition, Technology and Logistics identified the opportunity to adopt a common unmanned aircraft ground control station architecture that supports future capability upgrades through an open system and modular design. Similar to OSD's approach to ground control stations, the Air Force Unmanned Aircraft Systems Task Force expected future unmanned aircraft to be developed as open, modular systems to which new capabilities could be added instead of developing entirely new systems each time a new capability is needed.

DOD Continues to Increase Its Emphasis on and Funding For Unmanned Aircraft Systems

Since July 2009 when our report was issued, DOD has made several key investment decisions regarding unmanned aircraft systems that will likely impact those estimates. In general, these decisions reflect increased emphasis on developing more advanced unmanned aircraft capabilities and acquiring larger numbers of specific systems, but they do not appear to focus on increasing collaboration or commonality among systems.

The 2010 Quadrennial Defense Review (QDR) reported that "U.S. forces would be able to perform their missions more effectively—both in the near-term and against future adversaries—if they had more and better key enabling capabilities at their disposal." The QDR report included unmanned aircraft systems among these key enablers, and emphasized the importance of rapidly increasing the number and quality of unmanned aircraft systems—among other enablers—to prevail in today's wars, and to deter and defeat aggression in anti-access environments. The report also noted that: the Air Force is going to increase the total number of Predator/Reaper aircraft it plans to buy; the Army will accelerate the production of its Predator-class Sky Warrior[10] system; and the Navy will conduct field experiments with prototype versions of its Unmanned Combat Aircraft System, which, the QDR points out, offers the potential to greatly increase the range of strike, and intelligence, surveillance, and reconnaissance (ISR) operations from the Navy's carrier fleet.

---

[9]Pub. L. No. 110-417, § 144.

[10]The 2010 QDR specifically refers to the Extended Range Multi-Purpose system, which at the time of our 2009 report was being referred to as Sky Warrior.

As part of DOD's fiscal year 2011 budget development process, OSD made several unmanned aircraft-related adjustments to the services' budget submissions. As part of those adjustments, OSD:

- Directed the Army to stop development and initial fielding of its Fire Scout unmanned aircraft;
- Provided the Air Force an additional $344 million from FY2011 to FY2015 to develop, procure, and integrate counter-communication and counter-improvised explosive device jamming pods onto 33 MQ-9 Reaper aircraft, and directed the Air Force to present its assessment of platforms for this capability by June 1, 2010;
- Provided an additional $1.8 billion from FY2011 through FY2015 to purchase an additional 74 MQ-9 Reaper aircraft;
- Added $2 billion to the Navy budget from FY2013 to FY2015 to define requirements and develop unmanned carrier based capability, and directed the Navy to develop an execution plan by March 30, 2010;
- Added $201.6 million to the Global Hawk procurement budget to procure 19 Block 40 aircraft by 2015, and 22 total;
- Added $270.5 million for development and procurement of Global Hawk satellite communication terminals;
- Added $2.4 billion over the Future Years Defense Program to the Army's Extended Range Multi-Purpose (Sky Warrior) Aircraft budget to procure an additional 12 aircraft and 5 ground stations (one company) per year from 2011 through 2015.

In concert with the QDR and the fiscal year 2011 budget, DOD also published its first submission of a long-range, fixed-wing aviation procurement plan. Among other things, the plan addresses DOD's strategy for meeting the demand for persistent, unmanned, multi-role ISR capabilities by:

- Emphasizing "long-endurance, unmanned ISR assets—many with strike capabilities—to meet warfighter demands;
- Projecting an increase in the number of platforms in this category from approximately 300 in 2011 to more than 800 in 2020, nearly 200 percent increase;
- Noting the "replacement of Air Force Predators with more capable Reapers";
- Establishing a specific category for Unmanned Multi-role Surveillance and Strike systems, that distinguishes those systems from other types of aircraft, such as fighters and bombers;
- Noting that the department will continue to adapt the mix of unmanned and manned systems as security needs evolve; and

- Noting that unmanned systems are being considered as future long-range strike platforms and future fighter / attack aircraft.

Concluding Observations

In closing, recent experience in Iraq and Afghanistan has proven that unmanned aircraft are extremely valuable to the warfighter, and it is clear that more are needed. However, DOD will continue to be challenged to meet this increasing demand within available resources. Many of DOD's larger unmanned aircraft acquisition programs have experienced cost growth, schedule delays, and performance shortfalls, while not enough have achieved the efficiencies one might expect from commonality. DOD recognizes that to more effectively leverage its acquisition resources, it must achieve greater commonality among the military services' various unmanned system programs. However, in many cases the services have preferred to pursue unique solutions. In general, the military services continue to establish unique requirements and prioritize resources while foregoing opportunities to achieve greater efficiencies. As a result, commonality has largely been limited to system airframes, and in most cases, has not been achieved among payloads, subsystems, or ground control stations.

Opportunities for identifying commonality are greatest when requirements are being established. Therefore, as the department continues to develop and procure unmanned aircraft systems, it must take more care in setting requirements for those systems. Rather than looking for unique solutions to common problems, DOD must increasingly find common solutions to those problems. However, we recognize that commonality is not a panacea, and in some cases, given legitimate differences in operating environments or mission needs, may not make sense. We also recognize that achieving commonality is not always easy, especially given the strong service-driven acquisition processes and culture within the department. Therefore, in our July 2009 report we recommended that DOD (1) direct an objective, independent examination of unmanned aircraft requirements and report a strategy to Congress for achieving greater commonality among systems and subsystems, and (2) require future unmanned aircraft programs to take an open systems approach to product development and to clearly demonstrate that potential areas of commonality have been analyzed and identified. We believe that these steps could help overcome these barriers and could go a long way to ensuring that DOD maximizes efficiency as it continues to greatly increase emphasis on developing and acquiring more capable and larger quantities of unmanned aircraft.

Contacts and Staff Acknowledgments

For further questions about this statement please contact Michael J. Sullivan at (202) 512-4841. Individuals making key contributions to this statement include Bruce Fairbairn, Assistant Director; Travis Masters; Rae Ann Sapp; Leigh Ann Nally; Laura Jezewski; and Susan Neill.

Appendix I: Additional Unmanned Aircraft Program Data and Information

This appendix contains 3 tables that provide additional information about the 8 unmanned aircraft systems assessed in our July 2009 report. Table 4 contains the combined total development and procurement funding DOD has requested in its fiscal year 2011 budget submission for each of the programs. The budget data is presented in then year dollars and may not add precisely due to rounding. Tables 5 and 6 detail many of the key characteristics and compare the capabilities of the systems discussed in this statement.

Table 4: Fiscal Year 2011 Development and Procurement Funding Requested for Selected Unmanned Aircraft Programs

(Then year dollars in millions)

Aircraft	2010	2011	2012	2013	2014	2015	FY10-FY15
Reaper	$689.8	$1,474.3	$1,406.3	$1,600.2	$1,522.9	$1,661.1	$8,354.7
Global Hawk[a]	911.2	961.4	1,021.9	855.1	726.5	653.9	5,130.1
BAMS	439.0	529.3	541.0	744.5	807.2	723.2	3,783.9
Sky Warrior	568.2	644.2	544.2	519.9	532.2	497.4	3,306.1
Shadow	607.9	610.6	88.0	118.4	125.7	171.1	1,781.4
UCAS-D	304.9	266.4	216.0	165.2	51.3	52.7	1,056.4
Predator	188.9	208.2	123.0	99.7	75.1	44.8	829.5
Fire Scout	118.6	61.6	50.9	70.3	90.8	90.8	472.4
Total	$3,921.1	$4,781.4	$4,003.2	$4,178.1	$3,935.6	$3,895.1	$24,714.5

Source: DOD (data); GAO (analysis and presentation)

[a]Information on the RQ-4B Global Hawk is presented in this chart.

Table 5: Key Characteristics of Selected Unmanned Aircraft Systems

Aircraft	Length (feet)	Wing Span (feet)	Gross Weight (pounds)	Payload Capacity (pounds)	Endurance (hours)[a]	Maximum Altitude (feet)
Reaper	36	66	10,500	3,750	24	50,000
Global Hawk[b]	48	131	32,250	3,000	28	60,000
BAMS	48	131	32,250	3,200	34+	60,000
Sky Warrior	28	56	3,200	800	40	25,000
Shadow	11	14	375	60	6	15,000
UCAS-D	38	62	46,000	4,500	9	40,000
Predator	27	55	2,250	450	24+	25,000
Fire Scout	23	28	3,150	600	6+	20,000

Sources: DOD, *Unmanned Systems Roadmap 2007 – 2032* and BAMS Program Office

[a]Endurance capacity reported here is the maximum endurance possible, without external payloads. For some aircraft, the addition of external payloads can impact endurance capacity.

[b]Information on the RQ-4B Global Hawk is presented in this chart.

Table 6: Comparison of Key System Capabilities

	Imagery Intelligence			Signals Intelligence		
Aircraft	Electro-Optical / Infrared	Synthetic Aperture Radar	Full Motion Video	Communications Intelligence	Electronic Intelligence	Weapons
Global Hawk	X	X		X	X	
Predator	X	X	X	X		X
Reaper	X	X	X	X		X
Sky Warrior	X	X	X			X
Shadow	X		X			
Fire Scout - Navy	X		X			
Fire Scout - Army	X	X	X	X	X	
BAMS	X	X	X			

Source: DOD (data); GAO (analysis and presentation)

Note: While we also assessed the Navy's Unmanned Combat Aircraft System Demonstration (UCAS-D) as part of our review, UCAS-D is a demonstration effort and will not be equipped with any mission payloads.

This is a work of the U.S. government and is not subject to copyright protection in the United States. The published product may be reproduced and distributed in its entirety without further permission from GAO. However, because this work may contain copyrighted images or other material, permission from the copyright holder may be necessary if you wish to reproduce this material separately.

GAO's Mission	The Government Accountability Office, the audit, evaluation, and investigative arm of Congress, exists to support Congress in meeting its constitutional responsibilities and to help improve the performance and accountability of the federal government for the American people. GAO examines the use of public funds; evaluates federal programs and policies; and provides analyses, recommendations, and other assistance to help Congress make informed oversight, policy, and funding decisions. GAO's commitment to good government is reflected in its core values of accountability, integrity, and reliability.
Obtaining Copies of GAO Reports and Testimony	The fastest and easiest way to obtain copies of GAO documents at no cost is through GAO's Web site (www.gao.gov). Each weekday afternoon, GAO posts on its Web site newly released reports, testimony, and correspondence. To have GAO e-mail you a list of newly posted products, go to www.gao.gov and select "E-mail Updates."
Order by Phone	The price of each GAO publication reflects GAO's actual cost of production and distribution and depends on the number of pages in the publication and whether the publication is printed in color or black and white. Pricing and ordering information is posted on GAO's Web site, http://www.gao.gov/ordering.htm. Place orders by calling (202) 512-6000, toll free (866) 801-7077, or TDD (202) 512-2537. Orders may be paid for using American Express, Discover Card, MasterCard, Visa, check, or money order. Call for additional information.
To Report Fraud, Waste, and Abuse in Federal Programs	Contact: Web site: www.gao.gov/fraudnet/fraudnet.htm E-mail: fraudnet@gao.gov Automated answering system: (800) 424-5454 or (202) 512-7470
Congressional Relations	Ralph Dawn, Managing Director, dawnr@gao.gov, (202) 512-4400 U.S. Government Accountability Office, 441 G Street NW, Room 7125 Washington, DC 20548
Public Affairs	Chuck Young, Managing Director, youngc1@gao.gov, (202) 512-4800 U.S. Government Accountability Office, 441 G Street NW, Room 7149 Washington, DC 20548

STATEMENT

OF

PETER WARREN SINGER, PH.D.
SENIOR FELLOW AND DIRECTOR, 21st CENTURY DEFENSE INITITAIVE,
THE BROOKINGS INSTITUTION.

BEFORE

THE UNITED STATES HOUSE OF REPRESENTATIVES,
COMMITTEE ON OVERSIGHT AND GOVERNMENT REFORM,
SUBCOMMITTEE ON NATIONAL SECURITY AND FOREIGN AFFAIRS

Thank you very much Mr. Chairman and members of the committee for the opportunity to testify today. It is an honor to be a part of this important session on a topic that is crucial to our national security, but often crucially misunderstood.

As background, I work at the Brookings Institution, where I lead our research and analysis on 21st century defense issues, including on emerging actors and technologies in war. Several years back I became interested in just what was going on in this historic revolution, as robots begin to move into the fighting of our human wars. I interviewed everyone from robotic scientists and the science fiction writers who inspire them; 19 year old unmanned systems operators fighting wars 7,000 miles away, to those who command them, from Predator squadron commanders all the way up to the 4 star generals. I was interested in the politics of this and so met with White House advisors and military service secretaries. I was interested in the other side of perceptions and so pulsed the views of groups that ranged from Iraqi insurgents to Arab and Pakistani generals, and news editors across the Middle East and South Asia. Finally, I was interested in the questions of ethics, law, and right and wrong, and so interviewed military lawyers, as well as individuals with organizations like Human Rights Watch and the International Red Cross. Their stories, which I captured in my book *Wired for War*, were not just fascinating, but also shine a light on the political, legal, social, and ethical issues that ripple outwards.

What I would like to do today is briefly walk through where we stand now and then focus on what I see as some of the key policy questions that face us in this exciting, but sometimes scary new domain.

When the U.S. military went into Iraq in 2003, it only had a handful of unmanned systems in the air. The invasion force used zero unmanned ground vehicles. Today, we have over 7,000 of these unmanned systems in the air, ranging from 48-foot long Predators to micro-aerial vehicles that a single soldier can carry in their backpack. On the ground, we have over 12,000, such as the lawnmower-sized Packbot and Talon, which help find and defuse the deadly roadside bombs.

But we need to remember that while they often seem like they are straight out of science fiction, such PackBots and Predators are merely the first generation—the equivalent of the Model T Ford or the Wright Brothers' Flyer. Even more, they are being armed with everything from Hellfire missiles to 50 caliber machine guns. So, the term "killer app" (short for "killer applications," technologies that send massive bow waves onto industries, like what the I-Pod did to the music industry) is taking on an entirely new meaning.

The historic parallels that people make to where we stand now with robotics are instructive. Many scientists parallel unmanned systems today to where we were with "horseless carriages" back in 1909-1910, at the start of something so big we can only wrap our minds around what it is not. That is, automobiles and the resulting mechanization didn't just become change industry and warfare, it also reshaped our cities through the creation of suburbia, gave power to Middle East potentates who lived above oil deposits, and led to the requirement of new laws, "traffic laws."

Others, such as Bill Gates, have described robotics as being where computers were around 1980; if this is the case, think how the computer reshaped everything from our economy to our social relationships to how we fight wars and now even where we fight them (cyberwar). Finally, others make the parallel of robotics being much like the atomic bomb in the 1940s, a cutting-edge technology, of immense power and potential, but also a genie that we will not be able to put back into the box.

The point here is that every so often in history, the emergence of a new technology changes our world. Like gunpowder, the printing press, or even the atomic bomb, such "revolutionary" technologies are game-changers not merely because of their capabilities, but rather because the ripple effects that they have outwards onto everything from our wars to our politics. That is, something is revolutionary not so much because of what it can do, but rather the tough social, military, business, political, ethical, and legal questions it forces us to ask.

So, what are some of the key questions emerging in the growing field of robots and our wars?

1) Where is the (Unmanned) Military Headed?

The US military has gone from barely using robotics to using thousands of them in a bureaucratic blink of an eye. Its current plans, as one 3 star general described are that it will soon be using "tens of thousands." But as one USAF Captain put it to me out in CENTCOM, the problem is that "Its not "Let's think this better, it's only "Give me more.""

How do we ensure it buys the right ones and not over-priced, over-engineered, unwieldy systems that have gold-plated processors? How do we maintain competition and experimentation in an emerging sector in the defense industrial base? Knowing that having the right doctrine can be the difference between winning and losing wars, between committing America to the 21st century version of the Maginot Line vs. the Blitzkrieg, what are the proper organizational structures and doctrines for using these new systems? How do you ensure digital systems' security, so that foes can't tap into their communications, as insurgents in Iraq were able to do with a $30 software package they bought off the internet? How do we better support the men and women operating them, who may not be in the physical warzone, but are experiencing an entirely new type of combat stress? How do you ensure their future career prospects, so that organizational culture does not stymie change?

Another area is what is the proper division of warrior and civilian in this space? That is, if this area is the future of the force, is it proper that presently 75% of the maintenance and weapons loading of systems like the Predator have been outsourced to private contractors, including to controversial firms like Blackwater, while other Army systems operating in Iraq have been described as "government-owned-contractor operated?"

2) Are We Engaged in Three Wars?

As of March 12, 2010, American unmanned systems had carried out 118 known air strikes into Pakistan, well over double the amount we did with manned bombers in the opening round of the Kosovo War just a decade ago. By the old standards, this would be viewed as a war.

But why do we not view it as such? Is it because it is being run by the CIA, not by the military and thus not following the same lines of authority and authorization? Is it because Congress never debated it? Is it because we view the whole thing as costless (to us)? Or, are the definitions are changing, and what used to be war, isn't anymore?

3) What are the Perceptions of Robots in War?

How do robots change the public's and its representatives' relationship with war? Does the ability to YouTube video clips of combat turn war into a form of entertainment? Does it lead to Monday Morning Quarterbacking of our troops?

In turn, what about the perceptions of publics 7,000 miles away? Do they view our use of robots as "efficient" and "costless" as we report in our media, or as one newspaper editor described in Lebanon, "cruel and cowardly"? What does it mean when "drone" has become a colloquial word in Urdu and rock songs that Pakistani youth vibe to talk about America not fighting with honor? How does the reality of our painstaking efforts to act with precision emerge on the other side through a cloud of anger and misperceptions? Is America painting itself into the same corner that Israel did in Gaza, where it got very good at targeted strikes of Hamas leaders, but also good at unintentionally inducing 12 year old Palestinian boys to want to join Hamas?

4) Who Should be Allowed to Use This Technology?

It is not just the military that is using unmanned systems. DHS is flying them for border security. But so are some of the civilian vigilante "border militias," as well as criminals using them to scout targets. Local police departments like Miami Dade have gotten authorization to use them, and the FAA is exploring opening up the wider airspace, a crucial step to the continuation of the field. But, as one federal district court judge put it to me, the legal questions they raise in such areas as probable cause and privacy will likely reach to the Supreme Court. How about me, does the 2nd amendment cover my right to bear (robotic) arms? It sounds like a joke, but where does the line stop, and why?

5) Can the Laws Keep Up?

Robotics do not remove humans from the decision making, but they do move that human role geographically and chronologically. Decisions now made thousands of miles away, or even years ago, may have great relevance to a machine's actions (or inactions) in the here and now. But while technology moves at an exponential pace, our institutions are struggling to keep up. For example, the prevailing laws of war, the Geneva Conventions, were written in a year in which people listened to 45rpm records and the average home cost $7,400. Is it too much to ask them to regulate all the nuances of a 21st century technology like a Reaper system, that is being used to target an insurgent, who knows he is not supposed to hide out in a home surrounded by civilians, and that is exactly why he does? That is, with the 20th century laws under siege from both sides, do the laws need to be updated, how and in what ways?

6) Will America go the Way of Commodore Computers?

If this is a growing industry along the lines of computing or automobiles, why does the US not have a national robotics strategy, unlike many other states? If this field is also crucial to national security, how will America fare, especially given that 43 other countries are also building, buying, and using military robotics, including allies like the UK and Germany, but also states like Russia, China, and Iran? Can we stay ahead, or will we fall behind like so many other historic first-movers in technologic revolutions?

We may need to think even more broadly about this. In which direction does the state of the American manufacturing economy, as well as the state of science and mathematics education in our schools, have us headed? What does it mean for US security that the number of American students graduating each year with a degree in IT or engineering is slightly less than in 1986, but we have had a more than 500% rise in "parks, recreation, leisure and fitness studies"? What does it mean to have soldiers whose hardware increasingly says "Made in China" on the back and whose software increasingly is being written by someone in places like India?

7) What does the "Open Source" revolution hold for us?

Robotics are not like aircraft carriers or nuclear bombs; much of the technology is off-the shelf, and even do-it-yourself. Hitler's *Luftwaffe* may not have been able to fly across the Atlantic during World War II, but a 77 year old blind man has already done so with his own homemade drone. This technology will inevitably pass into the wrong hands, allowing small groups and even individuals to wield great power. Hezbollah flew four such weapons in its war with Israel.

As the 9-11 Commission warned, the tragedy that day was in part cause by a "failure of imagination." Can we apply the same lesson here? Can we develop a military and homeland-security strategy that considers not only how to use technology but how others will use it against us? That means widening the threat scenarios our agencies plan and train for, and the potential equipment they might need for a new range of defense. It also means new legal regimes to determine who should have access to such dangerous technologies—lest our best new weapon come back to bite us.

There are two summary points I would like to make about these questions.

The first is that within many of them we find the policy answers. That is, we may debate the specifics of the answer, but almost all extend from a gap of some sort in policy, as the technology races ahead of our institutions. The second is that these are all the sort of questions that used to be debated at science fiction conventions. But much like past technologies as the atomic bomb or the horseless carriage that were once just imaginary, they are now all too real. And thus they are crucial for serious people to engage upon.

Thank you very much for the opportunity to be part of this discussion today.

United States House of Representatives

Subcommittee on National Security

And Foreign Affairs

Statement prepared for the Hearing:

"Rise of the Drones:

Unmanned Systems and the Future of War"

Tuesday, 23 March 2010

Room 2154

Rayburn House Office Building

Statement of John Edward Jackson

Professor of Unmanned Systems

U.S. Naval War College

Newport, Rhode Island

This testimony reflects the personal views of the author and does not represent the official views of the Naval War College or the Department of the Navy.

Thank-you Mr. Chairman, and thank-you to the rest of the Subcommittee members for this opportunity to speak on two subjects about which I am passionate: The education of our dedicated warriors, and the role that unmanned systems can and should play in future military operations.

I am privileged to currently serve as a professor at the U.S. Naval War College in Newport, Rhode Island. In the fall of 1884 (just over 125 years ago) the College was formed as a place *"...of original research on all questions relating to war... or the prevention of war"*. At the time of the College's founding the flagship of the U.S. Navy's North Atlantic Squadron was *USS Tennessee*, a wooden-hulled steam-ship that also carried 22,000 square feet of sail as a back-up propulsion system! The young military officers who comprised the College's first class spent many long hours considering the ways in which evolving technologies, like wireless communications, electrical equipment, and long-range naval guns, would change the nature of warfare at the close of the 19th century.

Now... a century-and-a-quarter later, our students are still engaged in serious contemplation of the ways in which technology will alter the battlefield, this time in the form of a robotics revolution.

To be clear, the Naval War College is not a technical school, and issues of systems design and software architecture are better suited to the more junior officers attending the Naval Postgraduate School in Monterey, California where innovative research is being conducted at their Center for Autonomous Vehicle Research (CAVR). Rather, the mission of the Naval War College is to improve the ability of its students to make sound decisions in highly complex and stressful maritime and joint environments. If trends in computer science and robotics engineering continue, it is conceivable that autonomous systems could soon be developed that are capable of making life-and-death decisions without direct human intervention. The purpose of the new elective course entitled **"Unmanned Systems and Conflict in the 21st Century"** is to provide a forum for the consideration of the scientific, ethical and operational issues inherent in the employment of unmanned/robotic systems in the national security context.

The course provides the opportunity for students to study contemporary cases, trends, and issues in the development and use of unmanned systems in twenty-first century warfare. The students study and evaluate these systems from the tactical, operational, and strategic dimensions of war. In the course of their studies, they:

- Develop an appreciation for the current state of development in the field of unmanned/robotic systems in the air, ground and maritime domains.

- Understand the unique issues, opportunities, and challenges associated with the operational employment of unmanned/robotic systems.

- Appreciate the degree to which the use of unmanned/robotic systems could change the character of warfare in the 21st century.

- Assess the diverse ethical issues and attitudes relevant to the use of unmanned systems.

- Assess the unique leadership challenges that arise in the utilization of unmanned systems.

- Assess the diverse elements and key drivers affecting the decision-making process with regard to unmanned systems.

- Analyze the use of unmanned systems within the context of international law, the law of armed conflict, and the just war tradition.

In order to provide a more detailed overview of the course, a copy of the current syllabus is attached to this statement. In brief, the course looks at hardware issues in the air, land, and maritime environments and provides hands-on exposure to state-of-the-art systems. It then considers the issues of command and control, personnel manning, and the legal and ethical issues of employing these systems in national security situations. Students ultimately demonstrate their mastery of the subject through research requiring both written and verbal presentations (a sample of recent paper topics is attached). The course supports two specific Areas of Study (AOS): <u>Leadership and Ethics</u> and <u>Strategy, Operations, and Military History</u>.

It should be noted that significant support for the course has been provided by the Association for Unmanned Vehicle Systems International (AUVSI); a number of manufactures of unmanned systems; educational institutions including the Massachusetts Institute of Technology (MIT) and the U.S. Army War College; the Department of Defense; and Navy leaders and engineers from various program management offices and the Navy Undersea Warfare Center.

LESSONS LEARNED AND OBSERVATIONS

The following observations result from direct contact with the several dozen students who have taken the newly-established course during the current academic year, as well from discussions I have had with scores of military officers and other practitioners at meetings, symposia and conferences.

- I have found that military officers are generally well informed about the exponential growth in the use of unmanned systems throughout the Department of Defense; and they are highly motivated to probe beyond the headlines and promotional hype to ascertain the true capabilities and limitations of current technology.

- They have a keen interest in understanding the full range of research and development activities now underway, particularly with regard to those systems that could be fielded in the near-term that could impact on their critical war-fighting abilities.

- The intense desire for knowledge about unmanned systems is evident across all branches of the armed services, within many government agencies, and it extends to our international partners and allies.

- Students are acutely aware of the ethical and legal issues associated with the employment of robotic systems in combat. Of particular concern is the possibility that unmanned/robotic systems could be programmed to make lethal decisions in combat situations without active human participation in the "kill chain".

- They are keenly aware that unmanned/robotic systems could represent a true "revolution in military affairs" that has the potential to alter career fields, training pipelines, and combat tactics. They don't "fear" the future, but are mindful of the need to skillfully manage the impact of this disruptive technology.

My final observations pertain to the professionalism and vision of the many people I have encountered while developing and teaching this course. At Navy Headquarters the Chief of Naval Operations, Admiral Gary Roughead, has been a strong and vocal advocate for unmanned systems, about which he has said: *"This is the right way, this is where we have to go, and it will make us much, much more effective"*. (Remarks at Brookings Institution, 2 November 2009). I believe the message is getting through at all levels of the Navy, and whenever I have sought information, or requested senior leaders to travel to the College to speak with students, or when I have participated in conferences and symposia, I have received immediate and unqualified support. Additionally, I have been particularly impressed with the people I have met from academia, the scientific/engineering communities, and industry, all of whom are working tirelessly to bring the potential of unmanned systems to fruition. Finally, I salute our elected officials, as represented by the members of this subcommittee, who seek to ensure that neither organizational inertia nor the tendency to protect the status-quo will keep America from using the drive and genius of her people to devise and utilize technology and science as necessary to protect our citizens, our economy, and our nation.

John Edward Jackson jacksonj@usnwc.edu

Newport, Rhode Island March 2010

UNMANNED SYSTEMS AND CONFLICT IN THE 21ST CENTURY

Course Number SE-720

U.S. Naval War College

Newport, Rhode Island 02842

Prof. John E. Jackson
401-841-6515
jacksonj@usnwc.edu

Spring Trimester, 2009-2010

Member: Association for Unmanned Vehicle Systems International (AUVSI)

UNMANNED SYSTEMS AND CONFLICT IN THE 21ST CENTURY

1. Course Description:

The technological advances in the areas of computer science, artificial intelligence (AI) and robotics engineering achieved in the past decade have created the capacity for unmanned/robotic systems to move from the realm of science fiction onto the current battlefields of the 21st century. In 2009 there were over 5,300 unmanned aircraft systems in America's inventory, and the inventory of unmanned ground systems deployed to Iraq and Afghanistan exceeded 12,000. Sea-based unmanned/robotic systems are less fully developed, due in part to the demanding operating environment, but research and development is well underway on systems to be deployed in the air, on the surface, and under-seas. Critical design considerations for the Littoral Combat Ship, for example, have been made to accommodate the use of a variety of unmanned vehicles. One further indicator of Navy interest in this subject was the Chief of Naval Operations' October 2008 tasking to Strategic Studies Group XXVIII to study all aspects of the integration of unmanned systems into Navy force structure.

Many observers believe that the combination of super-computing technology and robotics engineering will drive changes within the military environment equal to the impact the wide-spread use of gunpowder had in the 16th/17th centuries and steam propulsion for ships had in the 19th and 20th centuries. These changes relate not only to the development and manufacture of highly-capable future systems, but also to issues regarding the ethics of their use, and the manner in which command and control will be exercised. The purpose of this course will be to acquaint students with the scientific, ethical and operational issues inherent in the employment of unmanned/robotic systems in the national security context.

2. Student Learning Outcomes:

This course provides the opportunity for students to study contemporary cases, trends, and issues in the use and development of unmanned systems in twenty-first century warfare. Students will study and evaluate these systems from the tactical, operational, and strategic dimensions of war looking at the multifaceted issues of their development and use. Graduates will be able to:

a. Develop an appreciation for the current state of development in the field of unmanned/robotic systems in the air, ground and sea domains.

b. Understand the unique issues, opportunities, and challenges associated with employment of unmanned/robotic systems.

c. Appreciate the degree to which the use of unmanned/robotic systems could change the nature of warfare in the 21st century.

d. Describe and assess the diverse ethical issues and attitudes in the use of unmanned systems.

e. Describe and assess the unique leadership challenges that arise in the utilization of unmanned systems.

f. Describe and assess the diverse elements and key drivers affecting the decision-making process with regard to unmanned systems.

g. Describe the use of unmanned systems within the context of international law, the law of armed conflict, and the just war tradition.

h. Describe the various legal issues and concerns with respect to the utilization of unmanned systems.

3. Methodology and Student Requirements:

Each student will complete a mini-research project requiring a written paper of 6-8 pages in length combined with a formal 10-15 minute presentation on a subject of their choice related to course themes. All work will be graded High Pass, Pass, or Fail. Successful completion of all course requirements will result in the award of two-hours of graduate credit.

4. Required Readings:

Wired for War: The Robotics Revolution and Conflict in the 21st Century by Dr. P.W. Singer, Penguin Press, New York, 2009.

Unmanned Systems Integrated Roadmap FY2009-2034, Department of Defense, Pentagon, Washington, D.C., 2009.

United States Air Force Unmanned Aircraft Systems Flight Plan 2009-2047, Headquarters, U.S. Air Force, Washington, D.C., 2009.

"Ethical and Legal Issues Associated with the Use of Unmanned/Robotic Systems" by Raul Pedroza, U.S. Naval War College, 2009.

5. Seminar and Reading Schedule:

Session 1:   Thursday, 11 March 2010

Title: Introduction to the Course

Session summary:

> This will be the kick-off session for the course. We will review the plan for the 10-weeks of
> instruction, and will make introductions.
>
> This session will also introduce the DOD Unmanned Systems Integrated Road Map, and the Air Force Unmanned Systems Flight Plan. A review of robotics in literature and cinema will also be conducted.

Readings:

- Wired, pp 19-93. Scan as much of entire book as possible prior to class.
- Roadmap, pp. xiii-15.
- Flight Plan, pp.14-19

Session 2:   Thursday, 18 March 2010

Title: Coming Soon to a Battlefield Near You: The Next Wave of WarBots

Session Summary:

> Part 1: Static display of Unmanned Undersea Systems from Bluefin Robotics.
>
> Part 2: VTC lecture by Brookings Senior Fellow Dr. P.W. Singer.
>
> We will spend the second half of the class in a VTC discussion with Dr. P.W. Singer, Senior Fellow at Brookings Institute, and author of the course's primary text Wired for War: The Robotics Revolution and Conflict in the 21st Century. Dr.

Singer will touch-upon many of the issues that will be addressed in subsequent sessions.

Readings:

- Wired, pp 109-123.

Session 3: Thursday, 25 March 2010

Title: Review and demonstration of unmanned/robotic air systems.

Guest lecturer #1: RDML Terry Kraft, USN, Director of ISR Capabilities, OPNAV Staff N2/N6

Guest Lecturer #2: MAJ Joe Campo, USAF, Predator pilot

Session summary:

This session will focus primarily on unmanned airborne systems being used by DOD and other government agencies.

Part 1: Static Display of ScanEagle and Hummingbird unmanned air systems courtesy of Boeing/Insitu.

Part 2: Briefing by RDML Terry Kraft on status of Navy UAS development.

Part 3: Briefing by MAJ Joe Campo about his experiences as a Predator pilot.

Readings:

- Wired, pp. 116-120.
- Roadmap, pp. 51-103.
- Flight Plan, scan pp 25-51.

Session 4: Thursday, 1 April 2010

Title: Review and demonstration of unmanned/robotic maritime systems.

Session summary:

This session will be conducted as a site-visit to the Navy Undersea Warfare Center (NUWC) Unmanned Systems Laboratory in Newport, Rhode Island.

Readings:

- Wired, pp. 114-116.

- Roadmap, 135-143.

- Other readings TBD

Session 5: Thursday, 8 April 2010

Title: Ethical and legal issues with the use of unmanned/robotic systems

Session summary:

Professors from the NWC International Law Department and the College of Operational and Strategic Leadership (Stockdale Chair of Military Ethics) will lead discussions of the many legal and ethical issues related to the use of unmanned systems.

Readings:

- Wired, pp. 413-427; pp. 382-412.

- Selected Reading: "Ethical and Legal Issues Associated with the use of Unmanned/Robotic Systems" by Raul Pedroza.

- Selected Reading: "The Predator War: What are the Risks of the CIA's Covert Drone Program?" by Jane Meyer, The New Yorker, October 26, 2009.

Session 6: Thursday, 15 April 2010

Title: Review and demonstration of unmanned/robotic ground systems.

Session Summary:

This session will focus on the various unmanned systems being used and/or developed to support ground operations. This will include a live demonstration of robotic systems provided by iRobot/Boeing and by Foster-Miller/QinetiQ Corporation.

Readings:

- Wired, pp. 110-114.
- Roadmap: pp 111-133.
- Other readings TBD

Session 7: Thursday, 29 April 2010

Title: Artificial Intelligence, How Much is Too Much?

Session summary: This session will investigate the state of development of artificially intelligent systems and will consider the potential for such systems to become superior to the human mind. We will discuss the "Terminator-syndrome" and the threat (if any) a malevolent AI system might create for humanity.

Session 8: Thursday, 6 May 2010

Title: Issues of Command and Control of unmanned/robotic systems

Session summary:

> This session will focus on issues of command and control, and will also review the CNO's Strategic Studies Group report on the integration of unmanned systems into naval operations in the 2025 and beyond timeframe.

Reading:

- Wired, pp. 123-134; pp. 205-236; pp. 344-359.
- Other readings TBD

Session 9: Thursday, 13 May 2010

Student Presentations on research topics.

Session 10: Thursday, 20 May 2010
Student Presentations and Course Wrap-up

Selected Bibliography

Arkin, Ronald C. *Governing Lethal Behavior in Autonomous Robots*. Boca Raton, FL: CRC Press, 2009

Axe, David. *Warbots*. Ann Arbor, MI: Nimble Books LLC. 2008.

Bar-Cohen, Yoseph and Hanson, David. *The Coming Robot Revolution*. New York: Springer Science and Business Media, Inc, 2009.

Bekey, George. *Robotics: State of the Art and Future Challenges*. London, England: Imperial College Press, 2008.

Krishnan, Armin. *Killer Robots: Legality and Ethicality of Autonomous Weapons*. Surrey, England: Ashgate Publishing, 2009.

Mets, David R. *Airpower and Technology: Smart and Unmanned Weapons*. Westport, CT: Praeger Security International, 2009.

Newcome, Laurence. *Unmanned Aviation: A Brief History of Unmanned Aerial Vehicles*. Reston, VA: American Institute of Aeronautics and Astronautics, 2004.

Wallach, Wendell and Allen, Colin. *Moral Machines: Teaching Robots Right from Wrong*. Oxford: Oxford University Press. 2009.

Yenne, Bill. *Attack of the Drones: A History on Unmanned Aerial Combat*. St. Paul, MN: Zenith Press. 2004.

Fiction

Asimov, Issac. *The Robot Series #1: The Caves of Steel*. New York: Bantam.1991.

_____. *The Robot Series #2*: *The Naked Sun*. New York: Bantam.1991.

_____. *The Robot Series #3*: *The Robots of Dawn*. New York: Bantam.1994.

_____, *The Robot Series #4: Robots and Empire*. London: Harper Collins.1996.

_____, *I, Robot*. New York: Bantam.1991.

Capek, Karel. *R.U.R: Rossum's Universal Robots*. New York: Penguin Books, 2004. (First published 1921).

Internet Resources

www.uvs-info.com — UVS International (Paris). A superb site with a tremendous amount of data. No charge, but must register for use. Sign-up for monthly "UVS News Flash".

www.auvsi.org — Site of the Association for Unmanned Vehicle Systems International. Many links available without joining, full resources available to registered members.

Sample Research Paper Topics: Winter Trimester 2009-2010

- "Bionics: Robotics in the Human Body"
- "Clausewitz vs. Unmanned Systems: When Theory and Technology Collide"
- "Defense Acquisition: Effects on the Future Unmanned Force Structure"
- "Do the Benefits of UCAS in a SEAD/DEAD Role Outweigh the Ethical Concerns?"
- "Gamers and Remotely Piloted Aircraft: Implications of the Video Game Generation in RPAs"
- "Is there a Requirement for Robotic Ethical and Moral Guidelines?"
- "Powering Unmanned Ground Vehicles"
- "Satellites and Remotely Piloted Vehicles: Two Remotely Operated Ships Passing In the Fight"
- "Swimming in Sensors, Drowning in Data"
- "Sensor Overload"
- "Tactical Application of Unmanned Systems"
- "The Singularity: The End of Humanity (NLT 2040)"
- "Unmanned Airships: Extreme Persistence"
- The Rise of the Machines: Ethics and Responsibility for Robots"
- "Unmanned Air Systems in Ballistic Missile Defense"
- "Unmanned Vehicles: A Viable Tool for the Marine Corps Logistician"

Statement of Michael S. Fagan
Chair, Unmanned Aircraft Systems (UAS)
Advocacy Committee
Association for Unmanned Vehicle Systems
International (AUVSI)

BEFORE THE HOUSE OVERSIGHT AND GOVERNMENT REFORM SUBCOMMITTEE ON NATIONAL SECURITY AND FOREIGN AFFAIRS

The Honorable John F. Tierney, Chairman
The Honorable Jeff Flake, Ranking Member

March 23, 2010

Mr. Chairman and members of the subcommittee, thank you for this opportunity to address the subcommittee. My name is Michael Fagan, and I Chair the Unmanned Aircraft Systems (UAS) Advocacy Committee for the Association for Unmanned Vehicle Systems International (AUVSI). It is an honor for me to be here representing the world's largest non-profit organization devoted exclusively to advancing the unmanned systems community.

While national defense still is the primary use of unmanned aircraft systems, there is much more that these systems can do (and are doing) to protect our nation and its citizens.

There are many technological reasons for the rise in the application of unmanned systems. I will briefly mention two. One reason is that detection, surveillance, measurement, and targeting are more effective when done as close to the observable as possible. This axiom applies to military systems as much as it does to everyday life. Small and medium-size UAS put military payloads close to hostile forces for very long periods of time while significantly reducing risk to friendly forces.

Another reason is that size, weight, and power (or SWAP) requirements for equivalent data processing and storage capabilities are decreasing. Last month, the Office of Naval Research completed the first test flights of key elements of a 50-pound persistent surveillance imagery payload for Shadow-class UAS. A similar operational payload is approximately 1,000 pounds and needs a commuter-size aircraft with crew to put it in its necessarily predictable orbit above the hostile target. As reduced SWAP allows more data processing to move onboard the UAS, available data link bandwidth can transmit to the ground more products that are more relevant to more analysts over larger areas – compared to raw data now sent to the ground. Additionally, processing onboard the unmanned aircraft automates intelligence-analysis tasks and increasingly permits the same number of analysts to be effective over a greater area.

UAS technology will continue to increase in the current U.S. regulatory environment, but it will more efficiently and effectively provide benefits to warfighters if UAS manufacturers can more easily and frequently get access to airspace that permits their research, development, test, and evaluation flights. AUVSI is in favor of FAA rulemaking that will enable increased airspace access for UAS manufacturers.

UAS manufacturers also depend significantly on engineers and scientists with relevant education. It is therefore equally important to national security that educational institutions with relevant science and engineering programs have routine access to national airspace. AUVSI is in favor of FAA rulemaking that permits educational institutions the airspace access they require to effectively educate the next generation in autonomous system technologies.

The wars in Iraq and Afghanistan have certainly driven demand for these systems, but many Americans are unaware that a ScanEagle UAS also aided in the successful recovery of Captain Phillips of the Maersk Alabama off the coast of Somalia last year. There are many other useful applications of unmanned technology -- air, ground and maritime systems -- that can protect our nation. Border patrol, emergency response, wildfire monitoring, civil unrest, search and rescue, law enforcement, port security, submarine detection and underwater mine-clearance, bulldozers for clearing land mines and IEDs, and ground robots used for explosive ordnance disposal are some examples of actual and potential robotic system missions for air, ground, and maritime systems.

Unmanned systems have been and will continue to be proven in war, and it is time to prove their heretofore under-recognized capabilities for increasing the effectiveness of civil law enforcement and public safety.

Technologies originally developed for warfare also must be transitioned to commercial operations. There is growing demand from the civil sector for UAS for uses such as precision farming, tracking shoals of fish, aerial photography, and more. This demand has the potential to drive a rapid advance of the technology. The United States has an opportunity to be at the forefront of the research and development of these advanced systems if it can address regulatory obstacles.

Our industry growth is adversely affected by International Traffic in Arms Regulations (ITAR) for export of certain UAS technologies, and by a lengthy license approval process by Political Military Defense Trade Controls (PM-DTC). AUVSI is an advocate for simplified export-control regulations and expedited license approvals for unmanned systems technologies.

Our hope is that today's hearing illuminates some of the ways that unmanned system technologies are changing and could change modern warfare, increase the safety of our men and women in the military, law enforcement, and public safety, and strengthen national security at all levels.

AUVSI's over 6,000 members from industry, government organizations, and academia are committed to fostering and promoting unmanned systems and related technologies.

Thank you for this opportunity to testify. I am happy to answer any questions you and Members of the Subcommittee may have.

UAS Facts, Figures and Quotes

UAS provide critical data for the Joint Warfighter - 24/7/365

- Predators and Reapers are providing more than 700 hours of full motion video every day (>22,000 hours per month) to the warfighter providing unmatched persistence and flexibility. Every second of every day, 40 Predator-series aircraft are airborne worldwide.

- Global Hawks are providing more than 550 hours of imagery each month.

- The U.S. Air Force's active and reserve components are fully engaged in UAS operations with 5 active MQ-1/9 squadrons; 1 active RQ-4 squadron; 2 Air Force Reserve MQ-1 squadrons; 4 ANG MQ-1 squadrons; 1 ANG MQ-9 squadron. The Air Force additionally is conducting UAS training operations at 5 Formal Training Units; 3 MQ-1 (1 ANG); 1 MQ-9; 1 RQ-4

The Air Force is currently flying 41 CAPS (28 Predator, 12 Reaper, 1 Global Hawk)

- The Air Force continues to grow its UAS combat capability and will add 6 MQ-1/9 CAPs over the next year. UAS growth will continue into FY11 with the culmination of 50 MQ-1/9 CAPs.

MQ-1 and RQ-4 aircraft are providing imagery in support of recovery operations in Haiti

- 1 RQ-4 CAP stood up 14 Jan 2010.
    -- CENTCOM asset redeployed enroute to the AOR to support SOUTHCOM.
    -- Over 50 hours of imagery provided to commanders and partner agencies.

USAF UAS Airframes

- Predators: 143 MQ-1 inventory, 78 in Combat
    - 690,000 hours flown; averaging 4,500 hrs/week

- Reapers: 40 MQ-9 in inventory, 19 in Combat
    - 95,000 hours flown; averaging 1,000 hrs/week

- Global Hawk: 18 RQ-4 in inventory (incl test), 7 combat coded, 4 deployed
    - 38,000 hours flown, averaging 120 hrs/week

Combat Flt Hours (FY08)
    Predator - 138,000
    Reaper - 12,770
    Global Hawk - 6,700

Combat Flt Hours (FY09)
    Predator – 179,000
    Reaper – 24,000
    Global Hawk – 6,500

Combat Flt Hours (FY10)
   Predator – 51,000
   Reaper – 12,000
   Global Hawk – 2,000

- Overall, use of Predator-series aircraft – In 2006 Predator-series aircraft flew 80,000 total flight hours. In 2009, this number more than tripled, growing to 295,000 flight hours for the year.

- Predator-series aircraft are expected to reach the one million flight hours milestone in late March or early April 2010. More than 85% of these hours have been accumulated in combat zones.

- In 2009, while performing border security missions, CBP flew more than 2000 hours and responded to more than 5,200 ground sensor activations; 2900 resulted in the identification of suspected illegal activity such as human or narcotics smuggling. These flight hours directly contributed to the seizure of more than 12,000 pounds of narcotics and the apprehension of more than 1500 suspects. (Source: DHS/CBP, March 2010)

Army UAS statistics as of 10 March 2010

- Approaching the 1 Million hours mark for all types of Army UAS
- 1,287 Fielded systems - in every Army Brigade Combat Team as organic equipment operated by line soldiers
- 291 Deployed systems in Iraq and Afghanistan, varies by density of units deployed - 15-17 per BCT, plus Military Police units
- Army Total Raven hours estimate: 275,269
- Army Raven Combat hours estimate: 241,775
- Over the last decade, the Army has experienced more than 4,200 percent growth in the operational tempo of unmanned aircraft systems.

Army Unmanned Aircraft Systems (UAS) are a rapidly growing capability that are quickly becoming indispensible to the Army. As an example of how quickly this capability has grown within the Army, when Operation Iraqi Freedom (OIF) began in March 2003, there were only 3 systems (13 aircraft) deployed in support of that operation. Today, we have 337 systems (1,013 aircraft) in OIF and Operation Enduring Freedom (OEF). This growth continues. For example, it took the Army 13 years to fly the first 100,000 hours of UAS. It took us less than a year to fly the next 100,000 hours, and now we fly more than 220,000 hours each year. By May 2010 Army UAS will have flown 1,000,000 flight hours, almost 90 percent of which were flown in support of combat operations.

Employment of UAS have become a critical part of how the Army conducts operations. The Army employs UAS across all echelons (to include Division/Corps level) providing dedicated/organic support for tactical maneuver and intelligence operations. Army UAS are predominately employed as tactical Reconnaissance, Surveillance, and Target Acquisition (RSTA) platforms supporting the Commander's scheme of maneuver. In this role, Army UAS have filled a critical need, providing "actionable" intelligence and decreasing the time between sensor and shooter (shortening the "kill chain").

Quotes from UAS Users

"The aircraft is an active contributor to the actual apprehension of these criminals. It's hard to imagine combat today without UAVs. The aircrafts' capabilities are continuously improving, and they are beginning to do a lot of the same missions as our manned aircraft."
    Maj. Jonathan Shaffner, brigade aviation officer and chief of operations, 2nd Stryker Brigade Combat Team, 25th Infantry Division

"The Shadow provides coverage for a lot of raids. We do road scans for roadside bombs and have actually caught terrorists in the act of implanting these bombs in the road. The UAV mission is imperative to today's combat operations. When the infantry troops are going into a certain area to clear buildings, we'll go in ahead of time and scan the area, and we'll be able to report to them exact grids of potential enemies in the area."
    Sgt. 1st Class David Norsworthy, a UAV platoon sergeant with 2nd BCT, 101st Abn. Div.

"In both Iraq and Afghanistan, my Marines prefer not to go outside the wire without first sending up a Raven or a Wasp to scan the area and see what's going on. We call the Raven and Wasp our Airborne Flying Binoculars and Guardian Angels."
    GySgt Butler, Infantry Patrol Sergeant, USMC, 2008

"The Shadow is the commander's eye on the battlefield. It's surveillance, target acquisition and route reconnaissance all in one "We saved countless lives, caught hundreds of bad guys and disabled tons of IEDs in our support of troops on the ground."
    Spc. Eric Myles, a UAV operator with Darkhorse Troop, 2nd Squadron, 1st Cavalry Regiment

"We've supported countless troops in contact. We've found IEDs and monitored vehicle-borne IEDs and supported a bunch of raids out there."
    Sgt. Reed Myers, a UAV operator with Darkhorse Troop, 2nd Squadron, 1st Cavalry Regiment

"We bring another level to the commanders overall situational awareness, allowing them a chance to check things that would take a lot longer for troops on the ground to find out: from seeing if a high-value insurgent is at his house to if smugglers are trying to cross a river or canal at night. If we see something suspicious, we report it. I personally found an improvised explosive device during Iron Pursuit, and I'm proud that I'm doing a job that can save lives of the troops on the ground."
    General comment by Spc. Sam Bishop, a UAV operator, 502nd Military Intelligence Company, 47th Forward Support Battalion, 2nd Brigade Combat Team, 1st Armored Division, Multi-National Division - Baghdad

"The simple fact is this technology saves lives."
    Sgt. 1st Class David Norsworthy, a UAV platoon sergeant with 2nd BCT, 101st Abn. Div.

Statement of
Dr. Edward Barrett
Director of Research, Stockdale Center
United States Naval Academy
"Rise of the Drones: Unmanned Systems and the Future of War"
House Committee on Oversight and Government Reform
Subcommittee on National Security and Foreign Affairs
March 23, 2010

Mr. Chairman and Subcommittee members, thank you for inviting me to speak about the ethical and educational ramifications of unmanned weapons systems. While ethical and educational inquiries often lag behind technological developments, the efforts of Mr. Singer and others have generated a timely and fruitful conversation among ethicists, educators, engineers, industry and military leaders, and civilian policymakers. Today's proceedings will surely contribute to that important conversation. Speaking as a civilian academic, then, I will first offer some reflections on these systems' ethical advantages and challenges, and then briefly discuss related educational initiatives at the Naval Academy.

The goals animating the development and use of unmanned platforms are ethically commendable. While sometimes excoriated as merely "prudential," effectiveness and efficiency are, fundamentally, moral imperatives. Constituted and supported by its citizen taxpayers, the liberal democratic state is morally obligated to effectively defend their human rights with their limited resources. Additionally, I would argue that unmanned systems are consistent with a society's duty to avoid unnecessary risks to its combatants—a duty that sparked the recent controversy over "Up-Armored" vehicles.

But these rights and corresponding duties must be weighed against other ethical considerations. The venerable just war criteria that now undergird international law specify both pre-war and in-war requirements. To be permissible, war must be the last resort available to a state intending to pursue a just cause, and circumstances must indicate a reasonable chance of succeeding in a proportionate manner. Once in war, harms must be necessary and proportionate. Vis-à-vis uninvolved civilians who maintain their rights *not* to be harmed, soldiers incur additional risk to avoid, and assign greater weight to, foreseeable harm to innocents.

In this ethical context, I want to highlight a few challenges generated by unmanned systems. First, they could encourage unjust wars. Cost reductions, of course, allow states to more readily pursue just causes. But favorable alterations to pre-war proportionality calculations could also reduce the rigor with which non-violent alternatives are pursued, and thus encourage unnecessary—and therefore unjust—wars. Additionally, and echoing concerns about private security firms and cyberattack capabilities, these less visible weapons could facilitate the circumvention of legitimate authority and pursuit of unjust causes. While these moral hazards obviously do not require us to maximize war costs and minimize unmanned systems, they do require efforts to better inform and monitor national security decisionmakers.

Second, once in war, remote-controlled systems—compared to manned—are said to induce unnecessary and disproportionate harm, especially to civilians. The argument assumes that soldiers engaged in such "virtual warfare" are less situationally aware, and also less restrained because of emotional detachment. However, accumulating data points in the opposite direction. Sensor improvements, lack of fear-induced haste, reduced anger levels, and crystal clarity about strike damage all combine to actually enhance awareness and restraint. If true, this data suggests that it would be unethical *not* to use remote-controlled systems—unless mitigating factors pertained.

This qualification brings us to a third ethical consideration. Reasonable chance of success in counterinsurgency and stability operations—where indigenous perceptions are crucial—requires the judicious use of unmanned systems. Mistaken perceptions that these weapons are less discriminate, or are indicative of flawed characters and/or tepid commitments, can undermine efforts unless accompanied by adjustments to footprints and perceptions. Also, ground robots are incapable of developing necessary personal relationships with local citizens. Again, these arguments suggest the need for prudent, not unreflective, limitations.

But the use of *autonomous* strike systems, my fourth and final ethical consideration, requires more caution. Again, effectiveness and efficiency would be important benefits. Truly robotic air, sea, and ground capabilities would sense, decide, and act more quickly than human beings. In an anti-access environment, a long range system capable of independently navigating to, identifying, and striking mobile targets would bolster conventional extended deterrence. And the need to merely monitor, not control, these systems would reduce personnel costs.

But exactly what would these autonomous systems sense, decide, and do? Would they adequately distinguish combatants from non-legitimate targets such as bystanding civilians and surrendering soldiers—a task complicated by counter-countermeasure requirements? Would they adequately—i.e., at least as well as humans—comply with necessity and proportionality imperatives? Minimizing these possible *in bello* errors would require the elusive ability to credibly attribute bad results to a culprit—designers, producers, acquisition personnel, commanders, users, and perhaps even robots themselves. And if the notion of "robot responsibility" ever becomes meaningful, would a self-conscious and willful machine choose its own ends, and even be considered a person with rights?

While robotic personhood is a titillating idea, nearer-term possibilities suggest a focus on the first few concerns. Computer scientist Ron Arkin is working assiduously to develop adequately discriminating and ethical robots with responsibility attribution capabilities, and I would not bet against him. But prior to that day, I would advise an incremental approach similar to that used with remote-controlled systems: intelligence missions first, strike missions later. Given the complexity involved, I would also restrict initial strike missions to non-lethal weapons and combatant-only areas. One possible exception to this non-lethal recommendation would involve autonomous systems targeting submarines, where one only would have to identify friendly combatants, enemy combatants, and perhaps whales.

In closing, I want to assure the Subcommittee that military educators are preparing military operators and staffers to think ethically about these and other emerging technologies. At the Naval Academy, the core ethics course taken by every second-year midshipman covers these issues and their theoretical foundations. Last year, Mr. Singer delivered an endowed lecture to the entire second year class. The Department of Leadership, Ethics, and Law offers an ethics elective dedicated to emerging military technologies, including robotics. History and engineering and courses that address these issues include History of Technology, Advanced Topics in Robotics, Advanced Technologies, Emerging Technologies, Principles of Systems Engineering, and Introduction to Systems Engineering. In April, 300 students in this last class will witness a debate between Ron Arkin and his less sanguine critic, Peter Asaro. And also in April, the Stockdale Center for Ethical Leadership, the Academy's ethics and military policy think tank, will host a two-day conference on the ethical ramifications of emerging military technologies attended by instructors from all U.S. service academies, staff colleges, and war colleges—and perhaps by a few congressional staffers.

Mr. Chairman and Subcommittee members, thank you for the opportunity to address these issues, and I look forward to your questions.

U.S. House of Representatives

Subcommittee on National Security and Foreign Affairs

Subcommittee Hearing:
"Rise of the Drones: Unmanned Systems and the Future of War"

Tuesday, March 23, 2010, 2:00 pm
2154 Rayburn House Office Building

Written Testimony Submitted By
Kenneth Anderson
March 18, 2010

Honorable Chairman and Members:

Introduction

1. My thanks to the Subcommittee, the Chairman and Members for inviting me to offer this testimony. My name is Kenneth Anderson. I am a professor of law at Washington College of Law, American University, Washington DC, and a member of the Hoover Task Force on National Security and Law, The Hoover Institution, Stanford University, Stanford CA. My areas of specialty include the laws of war and armed conflict, international law, and national security law. I have attached a brief biography as an appendix to this statement.

2. I have been invited to testify regarding the use and future of unmanned battlefield systems, and particularly unmanned aerial vehicles (UAVs) commonly referred to as "drones," in current and future US armed conflicts and uses of force. I focus my remarks on the legal policy implications of these systems, set in the framework of technological and strategic evolution.

3. The basic conclusions of my testimony are six:

   - First, the United States government urgently needs publicly to declare the legal rationale behind its use of drones, and defend that legal rationale in the international community, which is increasingly convinced that parts, if not all, of its use is a violation of international law.
   - Second, the legal rationale offered by the United States government needs to take account, not only of the use of drones on traditional battlefields by the US military, but also of the Obama administration's signature use of drones by the CIA in operations outside of traditionally conceived zones of armed conflict, whether in Pakistan, or further afield, in Somalia or Yemen or beyond. This legal rationale must be certain to protect, in plain and unmistakable language, the lawfulness of the CIA's participation in drone-related uses of force as it takes

place today, and to protect officials and personnel from moves, in the United States or abroad, to treat them as engaged in unlawful activity. It must also be broad enough to encompass the use of drones (under the statutory arrangements long set forth in United States domestic law) by covert civilian agents of the CIA, in operations in the future, involving future presidents, future conflicts, and future reasons for using force that have no relationship to the current situation.

- Third, the proper legal rationale for the use of force in drone operations in special, sometimes covert, operations outside of traditional zones of armed conflict is the customary international law doctrine of self-defense, rather than the narrower law of armed conflict.
- Fourth, Congress has vital roles to play here, mostly in asserting the legality of the use of drones. These include: *(i)* Plain assertion of the legality of the programs as currently used by the Obama administration, as a signal to courts in the US as well as the international community and other interested actors, that the two political branches are united on an issue of vital national security and foreign policy. *(ii)* Congressional oversight mechanisms should also be strengthened in ensuring Congress's meaningful knowledge and ability to make its views known. *(iii)* Congress also should consider legislation to clarify once and for all that that covert use of force is lawful under US law and international law of self-defense, and undertake legislation to make clear the legal protection of individual officers. *(iv)* Congress should also strongly encourage the administration to put a public position on the record. In my view, that public justification ought to be something (self-defense, in my view) that will ensure the availability of targeted killing for future administrations outside the context of conflict with Al Qaeda – and protect against its legal erosion by acquiescing or agreeing to interpretations of international law that would accept, even by implication, that targeted killing by the civilian CIA using drones is per se an unlawful act of extrajudicial execution.

The Multiple Strategic Uses of Drones and Their Legal Rationales

4. Seen through the lens of legal policy, drones as a mechanism for using force are evolving in several different strategic and technological directions, with different legal implications for their regulation and lawful use. From my conversations and research with various actors involved in drone warfare, the situation is a little bit like the blind men and the elephant – each sees only the part, including the legal regulation, that pertains to a particular kind of use, and assumes that it covers the whole. The whole, however, is more complicated and heterogeneous. They range from traditional tactical battlefield uses in overt war to covert strikes against non-state terrorist actors hidden in failed states, ungoverned, or hostile states in the world providing safe haven to terrorist groups. They include use by uniformed military in ordinary battle but also use by the covert civilian service.

5. Although well-known, perhaps it bears re-stating the when this discussion refers to drones and unmanned vehicle systems, the system is not "unmanned" in the sense that human beings are not in the decision or control loop. Rather, "unmanned" here refers solely to "remote-piloted," in which the pilot and weapons controllers are not

physically on board the aircraft. ("Autonomous" firing systems, in which machines might make decisions about the firing of weapons, raise entirely separate issues not covered by this discussion because they are not at issue in current debates over UAVs.)

6. *Drones on traditional battlefields.* The least legally complicated or controversial use of drones is on traditional battlefields, by the uniformed military, in ordinary and traditional roles of air power and air support. From the standpoint of military officers involved in such traditional operations in Afghanistan, for example, the use of drones is functionally identical to the use of missile fired from a standoff fighter plane that is many miles from the target and frequently over-the-horizon. Controllers of UAVs often have a much better idea of targeting than a pilot with limited input in the cockpit. From a legal standpoint, the use of a missile fired from a drone aircraft versus one fired from some remote platform with a human pilot makes no difference in battle as ordinarily understood. The legal rules for assessing the lawfulness of the target and anticipated collateral damage are identical.

7. *Drones used in Pakistan's border region.* Drones used as part of the on-going armed conflict in Afghanistan, in which the fighting has spilled over – by Taliban and Al Qaeda flight to safe havens, particularly – into neighboring areas of Pakistan likewise raise relatively few questions about their use, on the assumption that the armed conflict has spilled, as is often the case of armed conflict, across an international boundary. There are no doubt important international and diplomatic questions raised about the use of force across the border – and that is presumably one of the major reasons why the US and Pakistan have both preferred the use of drones by the CIA with a rather shredded fig leaf, as it were, of deniability, rather than US military presence on the ground in Pakistan. The legal questions are important, but (unless one takes the view that the use of force by the CIA is always and per se illegal under international law, even when treated as part of the armed forces of a state in what is unquestionably an armed conflict) there is nothing legally special about UAVs that would distinguish them from other standoff weapons platforms.

8. *Drones used in Pakistan outside of the border region.* The use of drones to target Al Qaeda and Taliban leadership *outside* of places in which it is factually plain that hostilities are underway begins to invoke the current legal debates over drone warfare. From a strategic standpoint, of course, the essence of much fighting against a raiding enemy is to deny it safe haven; as safe havens in the border regions are denied, then the enemy moves to deeper cover. The strategic rationale for targeting these leaders (certainly in the view of the Obama administration) is overwhelming. Within the United States, and even more without, arguments are underway as to whether Pakistan beyond the border regions into which overt fighting has spilled can justify reach to the law of armed conflict as a basis and justification for drone strikes.

9. *Drones used against Al Qaeda affiliates outside of AfPak – Somalia, Yemen or beyond.* The President, in several major addresses, has stressed that the United States will take the fight to the enemy, and pointedly included places that are outside of any

traditionally conceived zone of hostilities in Iraq or AfPak – Somalia and Yemen have each been specifically mentioned. And indeed, the US has undertaken uses of force in those places, either by means of drones or else by human agents. The Obama administration has made clear – entirely correctly, in my view – that it will deny safe haven to terrorists. As the president said in an address at West Point in fall 2009, we "cannot tolerate a safe-haven for terrorists whose location is known, and whose intentions are clear."[1] In this, the President follows the long-standing, traditional view of the US government endorsing, as then-State Department Legal Advisor Abraham Sofaer put it in a speech in 1989, the "right of a State to strike terrorists within the territory of another State where terrorists are using that territory as a location from which to launch terrorist attacks and where the State involved has failed to respond effectively to a demand that the attacks be stopped."[2]

10. The United States might assert in these cases that the armed conflict goes where the combatants go, in the case particularly of an armed conflict (with non-state actors) that is already acknowledged to be underway. In that case, those that it targets are, in its view, combats that can lawfully be targeted, subject to the usual armed conflict rules of collateral damage. One says this without knowing for certain whether this is, in fact, the US view – although the Obama administration is under pressure for failing to articulate a public legal view, this was equally the case for the preceding two administrations. In any case, however, that view is sharply contested as a legal matter. The three main contending legal views at this point are as follows:

    - One legal view (the traditional view and that presumably taken by the Obama administration, except that we do not know for certain, given its reticence) is that we are in an armed conflict. Wherever the enemy goes, we are entitled to follow and attack him as a combatant. Geography and location – important for diplomatic reasons and raising questions about the territorial integrity of states, true – are irrelevant to the question of whether it is lawful to target under the laws of war; the war goes where the combatant goes. We must do so consistent with the laws of war and attention to collateral damage, and other legal and diplomatic concerns would of course constrain us if, for example, the targets fled to London or Istanbul. But the fundamental right to attack a combatant, other things being equal, surely cannot be at issue.
    - A second legal view directly contradicts the first, and says that the legal rights of armed conflict are limited to a particular theatre of hostilities, not to wherever combatants might flee throughout the world. This creates a peculiar question as to how, lawfully, hostilities against a non-state actor might ever get underway. But the general legal policy response is that if there is no geographic constraint consisting of a "theatre" of hostilities, then the very special legal regime of the laws of armed conflict might suddenly, and without any warning, apply – and overturn – ordinary laws of human rights that prohibit extrajudicial execution, and certainly do not allow attacks subject merely to collateral damage rules, with

---

[1] Pres. Obama West Point speech.
[2] Abraham D. Sofaer, State Department Legal Adviser, [].

complete surprise and no order to it. Armed conflict is defined by its theatres of hostilities, on this view, as a mechanism for limiting the scope of war and, importantly, the reach of the laws of armed conflict insofar as the displace (with a lower standard of protection) ordinary human rights law. Again, this leaves a deep concern that this view, in effect, empowers the fleeing side, which can flee to some place where, to some extent, it is protected against attack.

- A third legal view (to which I subscribe) says that armed conflict under the laws of war, both treaty law of the Geneva Conventions and customary law, indeed accepts that non-international armed conflict is defined, and therefore limited by, the presence of persistent, sustained, intense hostilities. In that sense, then, an armed conflict to which the laws of war apply exists only in particular places where those conditions are met. That is not the end of the legal story, however. Armed conflict as defined under the Geneva Conventions (common articles 2 and 3) is not the only international law basis for governing the use of force. The international law of self-defense is a broader basis for the use of force in, paradoxically, more limited ways that do not rise to the sustained levels of fighting that legally define hostilities.

- Why is self-defense the appropriate legal doctrine for attacks taking place away from active hostilities? From a strategic perspective, a large reason for ordering a limited, pinprick, covert strike is in order to avoid, if possible, an escalation of the fighting to the level of overt intensity that would invoke the laws of war – the intent of the use of force is to *avoid* a wider war. Given that application of the laws of war, in other words, requires a certain level of sustained and intense hostilities, that is not always a good thing. It is often bad and precisely what covert action seeks to avoid. The legal basis for such an attack is not armed conflict as a formal legal matter – the fighting with a non-state actor does not rise to the sustained levels required under the law's threshold definition – but instead the law of self-defense.

- Is self-defense law simply a standardless license wantonly to kill? This invocation of self-defense law should not be construed as meaning that it is without limits or constraining standards. On the contrary, it is not standardless, even though it does not take on all the detailed provisions of the laws of war governing "overt" warfare, including the details of prison camp life and so on. It must conform to the customary law standards of necessity and proportionality – necessity in determining whom to target, and proportionality in considering collateral damage. The standards in those cases should essentially conform to military standards under the law of war, and in some cases the standards should be still higher.

11. The United States government seems, to judge by its lack of public statements, remarkably indifferent to the increasingly vehement and pronounced rejection of the first view, in particular, that the US can simply follow combatants anywhere and attack them. The issue is not simply collateral damage in places where no one had any reason to think there was a war underway; prominent voices in the international legal community question, at a minimum, the lawfulness of even attacking what they

regard as merely alleged terrorists. In the view of important voices in international law, the practice outside of a traditional battlefield is a violation of international human rights law guarantees against extrajudicial execution and, at bottom, is just simple murder. On this view, the US has a human rights obligation to seek to arrest and then charge under some law; it cannot simply launch missiles at those it says are its terrorist enemies. It shows increasing impatience with US government silence on this issue, and with the apparent – but quite undeclared – presumption that the armed conflict goes wherever the combatants go.

12. Thus, for example, the UN special rapporteur on extrajudicial execution, NYU law professor Philip Alston, has asked in increasingly strong terms that, at a minimum, the US government explain its legal rationales for targeted killing using drones. The American Civil Liberties Union in February 2010 filed an extensive FOIA request (since re-filed as a lawsuit), seeking information on the legal rationales (but including requests for many operational facts) for all parts of the drones programs, carefully delineating military battlefield programs and CIA programs outside of the ordinary theatres of hostilities. Others have gone much further than simply requests that the US declare its legal views and have condemned them as extrajudicial execution – as Amnesty International did with respect to one of the earliest uses of force by drones, the 2002 Yemen attack on Al Qaeda members. The addition of US citizens to the kill-or-capture list, under the authorization of the President, has raised the stakes still further. The stakes, in this case, are highly unlikely to involve President Obama or Vice-President Biden or senior Obama officials. They are far more likely to involve lower level agency counsel, at the CIA or NSC, who create the target lists and make determinations of lawful engagement in any particular circumstance. It is they who would most likely be investigated, indicted, or prosecuted in a foreign court as, the US should take careful note, has already happened to Israeli officials in connection with operations against Hamas. The reticence of the US government on this matter is frankly hard to justify, at this point; this is not a criticism per se of the Obama administration, because the George W. Bush and Clinton administrations were equally unforthcoming. But this is the Obama administration, and public silence on the legal legitimacy of targeted killings especially in places and ways that are not obviously by the military in obvious battlespaces is increasingly problematic.

13. *Drones used in future circumstances by future presidents against new non-state terrorists.* A government official with whom I once spoke about drones as used by the CIA to launch pinpoint attacks on targets in far-away places described them, in strategic terms, as the "lightest of the light cavalry." He noted that if terrorism, understood strategically, is a "raiding strategy" launched largely against "logistical" rather than "combat" targets – treating civilian and political will as a "logistical target" in this strategic sense – then how should we see drone attacks conducted in places like Somalia or Yemen or beyond? We should understand them, he said, as a "counter-raiding" strategy, aimed not at logistical targets, but instead at combat targets, the terrorists themselves. Although I do not regard this use of "combat" as a legal term – because, as suggested above, the proper legal frame for these strikes is

self-defense rather than "armed conflict" full-on – as a strategic description, this is apt.

14. This blunt description suggests, however, that it is a profound mistake to think that the importance of drones lies principally on the traditional battlefield, as a tactical support weapon, or even in the "spillover" areas of hostilities. In those situations, it is perhaps cheaper than the alternatives of manned systems, but is mostly a substitute for accepted and existing military capabilities. Drone attacks become genuinely special as a form of strategic, yet paradoxically discrete, air power outside of overt, ordinary, traditional hostilities – the farthest project of discrete force by the lightest of the light cavalry. As these capabilities develop in several different technological direction – on the one hand, smaller vehicles, more contained and limited kinetic weaponry, and improved sensors and, on the other hand, large-scale drone aircraft capable of going after infrastructure targets as the Israelis have done with their Heron UAVs – it is highly likely that they will become a weapon of choice for future presidents, future administrations, in future conflicts and circumstances of self-defense and vital national security of the United States. Not all the enemies of the United States, including transnational terrorists and non-state actors, will be Al Qaeda or the authors of 9/11. Future presidents will need these technologies and strategies – and will need to know that they have sound, publicly and firmly asserted legal defenses of their use, including both their use and their limits in law.

The Role of the CIA

15. The foregoing is intended to make clear that, first, "drone warfare" is really a set of heterogeneous activities, technologies, strategies, and actors. What the military does with drones in Afghanistan is different on many metrics from what the CIA does. The legal rationales offered to sustain the policy therefore need to take account of these differences as well.

16. The reality, however, is that the controversy centers on the use of drones that goes further "outwards" on the axes of (i) geographical and tactical remoteness of their use from a "traditional" armed conflict, (ii) the actor – uniformed military or civilian agency, (iii) covert or overt, or, in today's increasingly peculiar circumstances, "clandestine" – not covert, but not publicly acknowledged, either, (iv) relation to an existing overt war, or response to a new threat, thus raising the many controversies of "preventive" uses of force.

17. As a practical matter today, this simply means that what we are here discussing, ostensibly about "drones" and UAVs, is really, a millimeter below the surface, as much or more a discussion of the appropriate and lawful role of the CIA. We should be completely plain about this. Yes, there are issues related to the military use of drones on the battlefield. But the fundamental argument is over the expansion of drones beyond being a substitute weapon on traditional battlefields by the uniformed military to be a strategic tool used far from AfPak, by civilian agents of the CIA, even

perhaps – as Graham Allison and John Deutch urged in an op-ed last year – once again genuinely *covert*.

18. If I might respectfully suggest to the Subcommittee, then, the most fundamental question at issue here is not drones per se, but the technological development of drones forcing a discussion on the proper role of the CIA.

19. The lawfulness under US domestic law of the CIA to use force, in accordance with US statutes dating back to the founding of the CIA is not at issue. Use of force is not mentioned as such, but there is no question as to its lawfulness under US domestic law, provided that the steps required by the statute are taken. Congress has never seen fit overtly to name the use of force as such in the statutory language, preferring to use softer euphemisms and generalities. My view is that the time has come for Congress explicitly to revise the CIA statute to declare the so-called "Fifth Function" explicitly. I believe it is time to make that shift. Why?

20. Although unquestionably lawful under US domestic law, and viewed by the United States as lawful under international law as a matter of international law of self-defense, the international law position is beginning to come under pressure as parts of the international law community come to see human rights law, the laws of armed conflict, and other international law, as outlawing these kind of drone uses of force, particularly done covertly by civilian agents of a government on the territory of another. I emphasize that state practice and the views of states have long accepted the legitimacy (even without pronouncing on the legality of such) of such interventions, or at a minimum acquiesced in them, at least if they remain exceptional.

21. But that acquiescence by states as a matter of international law has largely concerned the issue of the territorial integrity of states set against an intervention aimed, for example, at attacking a terrorist in a safe haven. The practice today is contested increasingly on grounds of human rights – it is a prohibited act of extrajudicial execution, it is claimed, for the United States, for example, to launch its 2002 Yemen missile strike; it should have attempted, at a minimum, to detain and capture, offer surrender, before striking. And once having detained, it should then charge and try suspects on criminal grounds. That goes to a claim of unlawful targeting; in addition, of course, the concerns about unlawful collateral damage. The United States government, its agencies, officials, and counsel, in my view, have very little idea of the groundswell of an international campaign developing to de-legitimate the practice of drone warfare, starting with its conduct by the CIA.

22. Beyond this, as the CIA's central role in the Pakistan missions is on the front pages many days, important voices in the international law community are going further to attack not just the legal bases of drone warfare as such, but the fundamental premise of "intelligence" uses of force by the CIA. The view of much of the international law community is that all uses of force must be either law enforcement seeking to arrest a person, or else uniformed military of a state, engaged in armed conflict under its legal definitions in the laws of war. On that view, there is simply no legal space for the

CIA to undertake uses of force as it is doing in Pakistan or anywhere else. Armed conflict can only lawfully be undertaken by lawful combatants, and, on this view, officials of the CIA are not lawful combatants. Consider the following statement of a leading international law scholar at Notre Dame, Mary Ellen O'Connell:

> "Members of the CIA are not lawful combatants and their participation in killing persons – even in an armed conflict – is a crime."

23. This view was reinforced by a recent op-ed in the Washington Post by the eminent Georgetown and former West Point scholar of the laws of war, Gary Solis. Consider that Professor Solis said flatly that the CIA agents engaged in drone warfare are America's very own "unlawful combatants" – no less so, he said, than those they target:

> "In our current armed conflicts, there are two U.S. drone offensives. One is conducted by our armed forces, the other by the CIA. Every day, CIA agents and CIA contractors arm and pilot armed unmanned drones over combat zones in Afghanistan and Pakistan, including Pakistani tribal areas, to search out and kill Taliban and al-Qaeda fighters. In terms of international armed conflict, those CIA agents are, unlike their military counterparts but like the fighters they target, unlawful combatants. No less than their insurgent targets, they are fighters without uniforms or insignia, directly participating in hostilities, employing armed force contrary to the laws and customs of war. Even if they are sitting in Langley, the CIA pilots are civilians violating the requirement of distinction, a core concept of armed conflict, as they directly participate in hostilities."

24. My view differs from that of either Professor O'Connell or Professor Solis. In my view, the use of force by civilian CIA agents makes them combatants in the armed conflict underway in AfPak, because they are taking direct part in hostilities under traditional standards. In uses of force outside of AfPak, insofar as they are engaged in lawful exercises of the customary sovereign right of self-defense against a non-state actor, they are not combatants, but they are not thereby unlawful, nor is their use of force unlawful under international law. I have stated the basis for that legal conclusion in other places, and want to make a somewhat different observation here.

25. Professor Solis concludes by stating that the "prosecution of CIA personnel is certainly not suggested." I have trouble understanding why not, if one accepts the legal view of unlawful combatancy by the CIA. He has stated a case of legal equivalance between terrorist unlawful combatants and CIA unlawful combatants; why prosecution does not follow is unclear. Indeed, Professor O'Connell – saying aloud what others in the world of international law outside the United States, in my experience, think about this matter but do not quite so bluntly say – CIA participation in killing persons "is a crime." Professor Solis concludes by wondering whether

26. Drone warfare, therefore, raises questions on its own – but an underappreciated question is one that drone warfare *forces onto* the table – whether the United States government agrees, or does not agree, with its critics that the use of force by the CIA is unlawful combatancy, and a crime under international law. And if it disagrees as it presumably must, on what basis does it justify its lawfulness? Professor Solis suggests that senior officials of the CIA have understood that question; the inexplicable silence by the legal officials of the Obama administration as to how they would defend the international lawfulness of these policies suggests, to the contrary, that they do not. I would respectfully urge that Congress ought to insist on the appearance by the relevant legal counsel to agencies in the administration to state for the record why and on what basis these practices are lawful. Even as solely a practical matter, the silence of the administration's lawyers threatens to undermine the legitimacy, as a practical, moral, legal, and political matter, of targeted killing by the CIA using drones, in the future.

The Lack of Public Legal Justification from the Administration

27. The United States government is in a peculiar mismatch with respected to drone targeted killing, particularly as done by the CIA outside of immediate hostilities. On the one hand, senior leaders from the President and Vice-President on down, positively gush over the program and its successes. In today's newspapers – Thursday, March 11, 2010 – CIA director Panetta was on the front pages in what was clearly a carefully conceived effort to bolster the perception of the program in the public mind, defending its many successes. He is right to do so. The program, in my view, has been a stunning success. However, the CIA director's touting of its success is somewhat beside the point in current circumstances. The public, so far as I can tell, does not doubt the success of the program – it does precisely what the President said he would do, take the fight to the terrorists. The question is not its success – it is its *lawfulness*. And its lawfulness not as a single thing focused around drones, but instead the lawfulness of particular parts, conducted by particular actors.

28. As much as CIA director Panetta needs to put those successes to the public, he needs another public discussion entirely – one in which he puts his general counsel on stage to articulate why this form of killing people is not just effective, but legal. The utter failure of the administration's lawyers, anywhere across the administration, to do that is breathtaking. This reticence extends beyond the general counsel of the agency directly involved, the CIA; the principal US government lawyer on international law, the State Department Legal Adviser, has likewise expressed no formal view as to the correct legal view of targeted killing using drones by the CIA in a variety of settings. The mismatch can perhaps be best imagined by thinking of President Obama and Vice-President Biden and CIA director Panetta, standing in a press conference saying the glowing things they have said about these programs and celebrating their success – taking the fight to the enemy, denying them safe haven, going even into places like

Yemen or Somalia if necessary. Imagine, however, that their senior lawyers are standing beside them as they celebrate the ramping up of drone strikes month after month, far in excess of the Bush years – turning to the lawyers, all they can say is, "We have not yet reached a legal conclusion on this matter." Would the President and Vice-President have reason to believe they had been well-served by their lawyers? Lawyers, we all know, cannot be mere yes-men, and these issues are complicated and difficult – but we are more than a year into the Obama presidency, and these programs have emerged as expanding centerpieces of its on-offense counterterrorism policy, as well they should – and yet the lawyers publicly say nothing.

29. It is no doubt unfair to say that the lawyers have not reached conclusions. In some cases, that might be true, but most likely conclusions have been reached – but not shared with the public. This seems to me a profound mistake, on legal grounds and political grounds. There are ways to articulate the legal basis of these policies without having to reveal operational matters, and the legitimacy of these programs over the long term is distinctly at issue.

30. Congress could serve a useful function in pressing the administration to articulate publicly its rationale. Moreover, if Congress believes – as certainly I believe it should – that it ought to move legislatively to provide greater personal legal protections, against legal action both domestic and abroad, for CIA and other national security officers, then a crucial component of that is the public articulation of the basis on which the United States government will tell the rest of the world that its actions are lawful. That is not possible to do if all the relevant legal analysis is hidden away in a confidential OLC opinion.

"Reducing US Disincentives to Use Violence"

31. Many other issues could be considered in this discussion. The levels of collateral damage, for example – and whether they are high or low, to the extent they are known, on the basis of the realistic alternatives to targeted killing. Critics of drone-incurred collateral damage, after all, sometimes seem to imply that the alternative is no use of force at all – whereas a more realistic comparison might be the effects of a Pakistani army artillery barrage.

32. We could consider the evolution of technology and its likely effects on targeting decisions, collateral damage, ability to identify a target and get close enough to kill him and only him. I would urge Congress, in particular, to press forward research and development of these technologies, in part with funding, but also with assurances that those who develop these advances in far more discrete uses of force will not find themselves also at legal risk, domestically or abroad, down the road. The best is very firmly the enemy of the good – particularly when technology in these areas develops incrementally, one small step at a time.

33. I propose to close this written submission on a much more general note, however. It has become something of a trope in these discussions that the very availability of drones somehow makes it too easy for the US to kill, resort to violence and force. For example, I spoke before a group of US law students at one of our finest law schools some months ago, and was told by a student – in a group of students highly unlikely, it seemed to me, to enlist in the armed forces or join the CIA – that the problem with drones was that they "reduced the disincentives for the US to use force below their efficient level." I inquired as to how one would know the efficient level of the United States use of force and what would serve to induce it, and was told that the problem was that US personnel were not at personal risk – not enough US servicemen and women were at personal risk – from getting killed on the battlefield to deter the United States from needless violence. I was grateful, as I hope those reading this testimony would be, that one young woman spoke up, visibly upset, and said that it was hardly for privileged law students to sit and play God about efficient incentives and disincentives to violence – and she hoped that the United States would legally kill its enemies at least risk to its personnel.

34. I share that young woman's sentiment, of course. Drones are a major step forward toward much more discriminating uses of violence in war and self-defense – a step forward in humanitarian weapons technology. That development needs strong encouragement. But more fundamentally, I would hope that Congress would send strong signals, not just that it regards this technology as a humanitarian step forward, and not even the obvious message that the US intends to protect its serving men and women while it undertakes lawful uses of force and sees no contradiction with its legal duties in so doing.

35. The additional message that Congress should send is that targeted killing using drones has evolved as fast and far as it has since 9/11 because the United States confronts an enemy that has chosen to hide itself among civilians. What we call drone warfare is, as much as anything else, an attempt to counter, through technology, tactics by our enemies that rely upon systematic violations of the laws of war. The next time that someone raises the proposition that American "disincentives to violence" are reduced by drones, let them be reminded that, far more, drones represent an attempt to address an unlawful equilibrium in which one side takes obligations under the laws of war seriously, while the other side does not. That is the fundamental disequilibrium at work here, and drones the most measured and discrete response available – consistent with the policy that, as President Obama and all his administration have correctly said, the United States must take the fight to its enemies.

36. I thank the Subcommittee for its kind invitation to appear and offer testimony, and will endeavor to answer any questions you might have to the best of my ability.

Kenneth Anderson
Professor of Law
Washington College of Law, American University
Member, Task Force on National Security and Law

The Hoover Institution, Stanford University

Kenneth Anderson
Biography

Kenneth Anderson is a professor of law at Washington College of Law, American University, where he has taught since 1996. He is also a visiting fellow and member of the Hoover Institution Task Force on National Security and Law, Stanford University. Prior to joining the American University faculty, Mr. Anderson was general counsel to the Open Society Institute-Soros Foundations in New York City, and prior to that the director of the Human Rights Watch Arms Division. He is a 1986 graduate of Harvard Law School and 1983 graduate of the University of California, Los Angeles; he clerked in 1986-87 for Justice Joseph R. Grodin of the California Supreme Court. He is a member of the editorial board of the Journal of Terrorism and Political Violence, currently Treasurer and Executive Committee member of the Lieber Society of the American Society of International Law, and a blogger at Opinio Juris international law blog and the Volokh Conspiracy law blog. He is the author of numerous articles on international law and laws of war, and served as legal editor of Crimes of War (1998 Norton).

Congress of the United States
House of Representatives
Subcommittee on National Security and Foreign Affairs
Hearing: Rise of the Drones II: Examining the Legality of Unmanned Targeting
April 28, 2010

Lawful Use of Combat Drones

Mary Ellen O'Connell[*]

Combat drones are battlefield weapons. They fire missiles or drop bombs capable of inflicting very serious damage. Drones are not lawful for use outside combat zones. Outside such zones, police are the proper law enforcement agents and police are generally required to warn before using lethal force. Restricting drones to the battlefield is the most important single rule governing their use. Yet, the United States is failing to follow it more often than not. At the very time we are trying to win hearts and minds to respect the rule of law, we are ourselves failing to respect a very basic rule: remote weapons systems belong on the battlefield.[1]

I. A Lawful Battlefield Weapon

The United States first used weaponized drones during the combat in Afghanistan that began on October 7, 2001. We requested permission from Uzbekistan, which was then hosting the U.S. airbase where drones were kept.[2] We also used combat drones in the battles with Iraq's armed forces in the effort to topple Saddam Hussein's government that began in March 2003.[3] We are still using drones lawfully in the on-going combat in Afghanistan. Drones spare the lives of pilots, since the unmanned aerial vehicle is flown from a site far from the attack zone. If a drone is shot down, there is no loss of human life. Moreover, on the battlefield drones can be more protective of civilian lives than high aerial bombing or long-range artillery. Their cameras can pick up details about the presence of civilians. Drones can fly low and target more precisely using this

---

[*] Robert and Marion Short Chair in Law, University of Notre Dame.
[1] This testimony draws on the following publications by the author: *The Choice of Law Against Terrorism*, J. NAT'L SEC. LAW (forthcoming 2010); *Unlawful Killing with Combat Drones*, in SHOOTING TO KILL: THE LAW GOVERNING LETHAL FORCE IN CONTEXT (Simon Bronitt ed. Hart, forthcoming) available at http://ssrn.com/abstract=1501144; *Combatants and the Combat Zone*, 43 U. of RICH. L. REV. 845 (2009); *Enhancing the Status of Non-State Actors Through a Global War on Terror*, 43 COL. J. TRANS. LAW 435 (2005); *Ad Hoc War*, in KRISENSICHERUNG UND HUMANITÄRER SCHUTZ—CRISIS MANAGEMENT AND HUMANITARIAN PROTECTION 405 (2004).
[2] Eric Schmitt, *Threats and Responses: The Battlefield; U.S. Would Use Drones to Attack Iraqi Targets*, http://www.nytimes.com/2002/11/06/world/threats-responses-battlefield-us-would-use-dro...
[3] Brian M. Carney, *Air Combat by Remote Control*, May 12, 2008, http://online.wsj.com/article/SB121055519404984109.html#printMode

information. General McChrystal has wisely insisted on zero-tolerance for civilian deaths in Afghanistan. The use of drones can help us achieve that.

What drones cannot do is comply with police rules for the use of lethal force away from the battlefield. In law enforcement it must be possible to warn before using lethal force, in war-fighting this is not necessary, making the use of bombs and missiles lawful.

The United Nations Basic Principles for the Use of Force and Firearms by Law Enforcement Officials (*UN Basic Principles*) set out the international legal standard for the use of force by police:

> Law enforcement officials shall not use firearms against persons except in self-defense or defense of others against the imminent threat of death or serious injury, to prevent the perpetration of a particularly serious crime involving grave threat to life, to arrest a person presenting such a danger and resisting their authority, or to prevent his or her escape, and only when less extreme means are insufficient to achieve these objectives. In any event, intentional lethal use of firearms may only be made when strictly unavoidable in order to protect life.[4]

The United States has failed to follow these rules by using combat drones in places where no actual armed conflict was occurring or where the U.S. was not involved in the armed conflict.

On November 3, 2002, the CIA used a drone to fire laser-guided Hellfire missiles at a passenger vehicle traveling in a thinly populated region of Yemen. At that time, the Air Force controlled the entire drone fleet, but the Air Force rightly raised concerns about the legality of attacking in a place where there was no armed conflict. CIA agents based in Djibouti carried out the killing. All six passengers in the vehicle were killed, including an American.[5] In January 2003, the United Nations Commission on Human Rights received a report on the Yemen strike from its special rapporteur on extrajudicial, summary, or arbitrary killing. The rapporteur concluded that the strike constituted "a clear case of extrajudicial killing."[6]

Apparently, Yemen gave tacit consent for the strike. States cannot, however, give consent to a right they do not have. States may not use military force against individuals on their territory when law enforcement measures are appropriate. At the time of the strike, Yemen was not using military force anywhere on its territory. More recently, Yemen has been using military force to suppress militants in two parts of the country. The U.S.'s on-going drone use, however, has not been part of those campaigns.

---

[4] Adopted by the Eighth United Nations Congress on the Prevention of Crime and the Treatment of Offenders, Havana, Cuba, 27 August to 7 September 1990, http://www2ohchr.org/english/law/firearms.htm.
[5] Doyle McManus, *A U.S. License to Kill, a New Policy Permits the C.I.A. to Assassinate Terrorists, and Officials Say a Yemen Hit Went Perfectly. Others Worry About Next Time*, L.A. TIMES, Jan. 11, 2003, at A1.
[6] UN Doc. E/CN.4/003/3, paras. 37 – 39.

The United States has also used combat drones in Somalia probably starting in late 2006 during the Ethiopian invasion when the U.S. assisted Ethiopia in its attempt to install a new government in that volatile country. Ethiopia's effort had some support from the UN and the African Union. To the extent that the U.S. was assisting Ethiopia, our actions had some justification. It is clear, however, that the U.S. has used drone strikes independently of the attempt to restore order in Somalia. The U.S. has continued to target and kill individuals in Somalia following Ethiopia's pullout from the country.[7]

The U.S. use of drones in Pakistan has similar problems to the uses in Yemen and Somalia. Where military force *is* warranted to address internal violence, governments have widely resorted to the practice of inviting in another state to assist. This is the legal justification the U.S. cites for its use of military force today in Afghanistan and Iraq. Yet, the U.S. cannot point to invitations from Pakistan for most of its drone attacks. Indeed, for much of the period that the United States has used drones on the territory of Pakistan, there has been no armed conflict. Therefore, even express consent by Pakistan would not justify their use.

The United States has been carrying out drone attacks in Pakistan since 2004. Pakistani authorities only began to use major military force to suppress militancy in May 2009, in Buner Province. Some U.S. drone strikes have been coordinated with Islamabad's efforts, but some have not. Some strikes have apparently even targeted groups allied with Islamabad.

II. The Battlefield Defined

The Bush administration justified the 2002 Yemen strike and others as justified under the law of armed conflict in the "Global War on Terror."[8] The current State Department Legal Adviser, Harold Koh, has rejected the term "Global War on Terror", preferring to base our actions on the view that the U.S. is in an "armed conflict with al-Qaeda, the Taliban and associated forces."[9] Under the new label, the U.S. is carrying out many of the same actions as the Bush administration under the old one: using lethal force without warning, far from any actual battlefield.

Armed conflict, however, is a real thing. The United States is currently engaged in an armed conflict in Afghanistan. The United States has tens of thousands of highly trained troops fighting battles with a well-organized opponent that is able to hold territory. The

---

[7] S. Bloomfield, *Somalia: The World's Forgotten Catastrophe*, THE INDEPENDENT, 9 February 2008, available at http://www.independent.co.uk/news/world/africa/somalia-the-worlds-forgotten-catastrophe-778225.html (last visited June 8, 2009); *See also*, CBS NEWS, *U.S. Missile Strike Hits Town in Somalia*, 3 March 2008, available at http://www.cbsnews.com/stories/2008/03/03/world/main3898799.shtml (last visited June 8, 2009); SPIEGEL ONLINE, *A Strike Against Al-Qaeda's Hornet's Nest*, 1 September 2007, available at http://www.spiegel.de/international/0,1518,458597,00.html (last visited June 8, 2009).
[8] Remarks of National Security Adviser Condoleeza Rice, FOX NEWS Sunday with Tony Snow (*Fox News* v. broadcast, Nov. 10, 2002) transcript *available at* 2002 WL 7898884.
[9] Harold Hongju Koh, *The Obama Administration and International Law*, Annual Meeting of the American Society of International Law, Washington, D.C., March 25, 2010, www.state.gov.

situation in Afghanistan today conforms to the definition of armed conflict in international law. The International Law Association's Committee on the Use of Force issued a report in 2008 confirming the basic characteristics of all armed conflict: 1.) the presence of organized armed groups that are 2.) engaged in intense inter-group fighting.[10] The fighting or hostilities of an armed conflict occurs within limited zones, referred to as combat or conflict zones. It is only in such zones that killing enemy combatants or those taking a direct part in hostilities is permissible.

Because armed conflict requires a certain intensity of fighting, the isolated terrorist attack, regardless of how serious the consequences, is not an armed conflict. Terrorism is crime. Members of al Qaeda or other terrorist groups are active in Canada, France, Germany, Indonesia, Morocco, Saudi Arabia, Spain, the United Kingdom, Yemen and elsewhere. Still, these countries do not consider themselves in a war with al Qaeda. In the words of a leading expert on the law of armed conflict, the British Judge on the International Court of Justice, Sir Christopher Greenwood:

> In the language of international law there is no basis for speaking of a war on Al-Qaeda or any other terrorist group, for such a group cannot be a belligerent, it is merely a band of criminals, and to treat it as anything else risks distorting the law while giving that group a status which to some implies a degree of legitimacy.[11]

To label terrorists "enemy combatants" lifts them out of the status of *criminal* to that of *combatant*, the same category as America's own troops on the battlefield. This move to label terrorists combatants is contrary to strong historic trends. From earliest times, governments have struggled to prevent their enemies from approaching a status of equality. Even governments on the verge of collapse due to the pressure of a rebel advance have vehemently denied that the violence inflicted by their enemies was anything but criminal violence. Governments fear the psychological and legal advantages to opponents of calling them "combatants" and their struggle a "war."

President Ronald Reagan strongly opposed labeling terrorists combatants. He said that to "grant combatant status to irregular forces even if they do not satisfy the traditional requirements ... would endanger civilians among whom terrorists and other irregulars attempt to conceal themselves."[12]

The United Kingdom and other allies take the same position as President Reagan: "It is the understanding of the United Kingdom that the term 'armed conflict' of itself and in its

---

[10] International Law Association, Initial Report of the Use of Force Committee, The Meaning of Armed Conflict in International Law (Aug. 2008, Rio de Janeiro) www.ilahq.org. *See also*, Mary Ellen O'Connell, *Defining Armed Conflict*, 13 J. CONFLICT & SECURITY LAW (2008).

[11] Christopher Greenwood, *War, Terrorism and International Law*, 56 CURR. LEG. PROBS. 505, 529 (2004).

[12] Ronald Reagan, Letter of Transmittal, The White House, January 29, 1987, *reprinted in* 81 AM. J. INT'L L. 910 (1987); but see response, Hans Peter Gasser, *An Appeal for Ratification by the United States*, 81 AM. J. INT'L L. 912 (1987).

context denotes a situation of a kind which is not constituted by the commission of ordinary crimes including acts of terrorism whether concerted or in isolation."[13]

In the United States and other countries plagued by al Qaeda, institutions are functioning normally. No one has declared martial law. The International Committee of the Red Cross is not active. Criminal trials of suspected terrorists are being held in regular criminal courts. The police use lethal force only in situations of necessity. The U.S.'s actions today are generally consistent with its long-term policy of separating acts of terrorism from armed conflict—except when it comes to drones.

III. Battlefield Restraints

Even when the U.S. is using drones at the request of Pakistan in battles it is waging, we are failing to follow important battlefield rules. The U.S. must respect the principles of necessity, proportionality and humanity in carrying out drone attacks. "Necessity" refers to military necessity, and the obligation that force is used only if necessary to accomplish a reasonable military objective.[14] "Proportionality" prohibits that "which may be expected to cause incidental loss of civilian life, injury to civilians, damage to civilian objects, or a combination thereof, which would be excessive in relation to concrete and direct military advantage anticipated."[15] These limitations on permissible force extend to both the quantity of force used and the geographic scope of its use.

Far from suppressing militancy in Pakistan, drone attacks are fueling the interest in fighting against the United States. This impact makes the use of drones difficult to justify under the terms of military necessity. Most serious of all, perhaps, is the disproportionate impact of drone attacks. A principle that provides context for all decisions in armed conflict is the principle of humanity. The principle of humanity supports decisions in favor of sparing life and avoiding destruction in close cases under either the principles of necessity or proportionality. According the International Committee of the Red Cross, the principles of necessity and humanity are particularly important in situations such as Pakistan:

> In classic large-scale confrontations between well-equipped and organized armed forces or groups, the principles of military necessity and of humanity are unlikely to restrict the use of force against legitimate military targets beyond what is already required by specific provisions of IHL. The practical importance of their restraining function will increase with the ability of a party to the conflict to control the circumstances and

---

[13] Marco Sassòli, *Use and Abuse of the Laws of War in the "War on Terrorism,"* 22 LAW & INEQ. 195, (2004), citing Reservation by the United Kingdom to Art. 1, para. 4 & Art. 96, para. 3 of Protocol I.

[14] W. Michael Reisman & Douglas Stevick, *The Applicability of International Law Standards to United Nations Economic Sanctions Programmes*, 9 EUR. J. INT'L L. 86, 94-95 (1998).

[15] Additional Protocol I, Art. 51(5); *see also* Judith Gardam, *Proportionality and Force in International Law*, 87 AM. J. INT'L L. 391 (1993).

> area in which its military operations are conducted, may become decisive where armed forces operate against selected individuals in situations comparable to peacetime policing. In practice, such considerations are likely to become particularly relevant where a party to the conflict exercises effective territorial control, most notably in occupied territories and non-international armed conflicts.[16]

Another issue in drone use is the fact that strikes are carried out in Pakistan by the CIA and civilian contractors. Only members of the United States armed forces have the combatant's privilege to use lethal force without facing prosecution. CIA operatives are not trained in the law of armed conflict.[17] They are not bound by the Uniform Code of Military Justice to respect the laws and customs of war. They are not subject to the military chain of command. This fact became abundantly clear during the revelation of U.S. use of harsh interrogation tactics. Given the negative impact of that unlawful conduct on America's standing in the world and our ability to promote the rule of law, it is difficult to fathom why the Obama administration is using the CIA to carry out drone attacks, let alone civilian contractors.

Conclusion

The use of military force in counter-terrorism operations has been counter-productive. Military force is a blunt instrument. Inevitably unintended victims are the result of almost any military action. Drone attacks in Pakistan have resulted in large numbers of deaths and are generally seen as fueling terrorism, not abating it. In Congressional testimony in March 2009, counter-terrorism expert, David Kilcullen, said drones in Pakistan are giving "rise to a feeling of anger that coalesces the population around the extremists and leads to spikes of extremism well outside the parts of the country where we are mounting those attacks."[18] Another expert told the *New York Times*, "The more the drone campaign works, the more it fails—as increased attacks only make the Pakistanis angrier at the collateral damage and sustained violation of their sovereignty.'"[19] A National Public Radio Report on April 26, 2010, pointed out that al Qaeda is losing support in the Muslim world because of its violent, lawless tactics.[20] We can help eliminate the last of that support by distinguishing ourselves through commitment to the rule of law, especially by strict compliance with the rules governing lethal force.

---

[16] International Committee of the Red Cross, *Interpretative Guidance on the Notion of Direct Participation in Hostilities under International Humanitarian Law* 27 (May 2009), at 80-81.
[17] William C. Banks, expert on U.S. national security law, e-mail to the author, Sept. 28, 2009 (on-file with the author).
[18] Hearing of the House Armed Services Committee, Effective Counterinsurgency: the Future of the U.S. Pakistan Military Partnership, April 23, 2009 (Testimony of David Kilcullen).
[19] Jan. 23, 2010, p. A3
[20] Dina Temple-Raston, *As Support Fades, Al-Qaida Shows Signs of Decline*, NPR, April 26, 2010, www.npr.org.

Statement of
David W. Glazier
Professor of Law
Loyola Law School Los Angeles

Hearing on
RISE OF THE DRONES II:
EXAMINING THE LEGALITY OF UNMANNED TARGETING

United States House of Representatives
Committee on Oversight and Government Reform
Subcommittee on National Security and Foreign Affairs
April 28, 2010

Introduction

I would like to express my sincere appreciation to the Chairman and members of the Committee for the opportunity to address this timely and important subject.

My name is David Glazier and I am a professor of law at Loyola Law School Los Angeles. I spent twenty-one years as a U.S. Navy surface warfare officer, culminating in command of USS George Philip (FFG 12), before retiring to attend law school at the University of Virginia in the summer of 2001. I began detailed research on law of war related topics during the spring of my first year of law school which I have continued to the present date. I remained in Charlottesville as a research fellow at the Center for National Security Law for two years following my law school graduation, publishing law review articles on law of war topics and developing and teaching new Virginia course offerings on the Law of War and American Military Justice. I started work at Loyola in 2006 where I continue to teach and write about the Law of War as well as U.S. Constitutional Law, Foreign Relations Law, and International Law.

Witnesses during last month's hearing before this subcommittee noted that the use of unmanned drones gives rise to a number of potential technological, ethical, and legal issues. Other recent public discussion has questioned the wisdom of their current employment in Afghanistan and Pakistan from a policy perspective, suggesting that attacks perceived as indiscriminate because of "collateral" civilian casualties may do more harm in fueling support for our adversaries than we gain from these strikes. It is my intention to limit my remarks to what I understand to be the focus of this hearing – legal issues implicated by drone use – and I will concentrate on my own area of expertise, the law of war. I would just note upfront that while law of war compliance is not a panacea, past American military and political leaders, dating from George Washington, have perceived policy and public relations advantages from faithful law of war compliance even when not reciprocated by our adversaries. A number of these same officials played a substantial role in developing the current rules.

Although previous government use of the incoherent "war on terror" nomenclature has likely contributed to some of the confusion on this subject, I do not see any credible basis to dispute the idea that the United States is lawfully engaged in an armed conflict against al Qaeda and the resurgent Taliban in which it has the right to draw legal authority from the law of war. The world community has generally recognized the events of "9/11" as constituting an armed attack allowing U.S. invocation of the right of self-defense under Article 51 of the United Nations Charter. 9/11 remains the only event that the liberal western democracies comprising the North Atlantic Treaty Organization have recognized as an armed attack on a member state. Even more importantly from a domestic law perspective, Congress exercised its constitutional authority to permit a military response in the September 2001 Authorization for the Use of Military Force (AUMF). Leading foreign relations scholars, and more importantly, the U.S. Supreme Court in its *Hamdi* and *Hamdan* decisions, have recognized the AUMF as the functional equivalent of a declaration of war, permitting the President to exercise authority granted by the law of war in countering those responsible for perpetrating or aiding and abetting the 9/11 attacks. The fact that al Qaeda is not an actual state does not bar this from being an armed conflict. The United States has fought a number of previous conflicts against groups it refused to recognize as legitimate nations, including the Confederacy, Indian tribes, Philippine Insurgents, and the Viet Cong while applying law of war rules to both sides.

The Law of War and the Lawfulness of Drone Weapons

The necessary starting point in evaluating the legality of any weapons use is identifying the scope and content of the contemporary law of war. The only part of this corpus juris most Americans seem to have heard of is the four Geneva Conventions of 1949, although in my experience few even among educated audiences really know what they say. The irony is that the Conventions are narrowly focused on providing protections for persons who either never were, or no longer are, legitimate objects of attack including the sick, wounded, shipwrecked, prisoners of war, and civilians actually in the hands of a foreign belligerent. They thus have virtually nothing to say about the actual conduct of hostilities. It is necessary to look to the larger, but much less well known, body of other treaties and customary law rules governing armed conflict to find authority relevant to the question before this subcommittee. It is also worth noting that this law includes both specific substantive rules and more general governing principles. The latter, such as the requirement to distinguish between lawful military objects of attack and protected civilian persons and objects, remain fully applicable regardless of changing circumstances and can readily be applied even to technological innovations unforeseeable at the time of the rules' development. In my opinion most of the frequently heard assertions that the law of war requires urgent updating to address "new" situations or technologies are overstated or outright erroneous.

Evaluating the legality of a weapons system requires a multi-part analysis. First, many types of weapons are the subject of specific prohibitions found either in treaty law, such as bans on chemical and biological weapons, blinding lasers, and anti-personnel land mines, or customary law rules, such as prohibitions on any use of exploding anti-personnel bullets, poisoned weapons, serrated bayonets, etc. Treaty bans are legally applicable only to states actually party to the relevant accord while customary law prohibitions are considered binding on all nations. There are no specific law of war bans which would apply to drone use, however.

Assuming a weapon falls outside these specific prohibitions, it is then necessary to evaluate its conformance with several customary law principles. As previously mentioned, the law of war requires "distinction" between "military objects" which are lawful objects of attack,[*] and civilian objects which are not. A weapon incapable of discriminate employment, such as the V-1 flying bombs Germany fielded late in World War II, incapable of striking anything more precise than a general sector of a target city, runs afoul of this rule and should be considered unlawful. While a drone launching missiles can be misused as readily as any other weapons system, these weapons capable of discriminate employment rivaling or exceeding virtually any other system in any nation's inventory. Indeed, it has been suggested that remote drone pilots, freed from the stress factor of being physically at risk while operating these systems, are able to be more careful in distinguishing targets than traditional warrior counterparts within the range of enemy weapons.

A second relevant principle bans the use of any weapon which causes "superfluous injury" or "unnecessary suffering." While these terms are subject to some interpretation, it is indisputable that the law of war permits the killing or injuring of opposing combatants by means of both kinetic and explosive projectiles and missiles, so a drone launching weapons equivalent to those traditionally employed from other platforms is not at risk of violating this rule.

Some concern has been voiced about the remote nature of drone combat, particularly as currently employed in which operators may be shift-workers located in the United States, literally oceans away from the scene of combat and effectively invulnerable to counterattack. This practice violates historical notions of chivalry which emphasized personal combat—whether between knights of medieval Europe or flying aces over the trenches of World War I battlefields. Although commentators often identify chivalric values as contributing to modern law of war development, there is nothing in the law that requires mutual exposure to risk between opposing forces. Indeed, the history of lawful weapons developments reveals a continuing effort by virtually all states to develop weapons more effective or longer-ranged than those of potential adversaries, allowing the achievement of maximum military advantage at minimum risk to their own forces. The replacement of swords, spears, and bows with firearms, the development of artillery systems of increasing range and accuracy, and the torpedo-equipped submarine are all examples of lawful past efforts to achieve relative (albeit ultimately temporary) invulnerability. Naval and land mines are historic examples of lawful early efforts to use robotic technologies to inflict military damage in remote locations.

Legal Considerations in Drone Employment

Although there is nothing unlawful per se about using drones in combat, it is still necessary that their use to conform to traditional rules governing the conduct of armed conflict, commonly known as *jus in bello*. These rules define, inter alia, who and what may be targeted, as well as when and where attacks may be conducted, and who may conduct them. While drones may be employed in a manner fully consistent with the law of war, there are some problematic aspects of

---

[*] A military object may be one directly associated with the armed forces, such as a military base, ship, tank, aircraft, etc. or it may be a otherwise civilian object which is being used to achieve a specific military advantage, such as railroad being used to move troops or resupply combat units or a highway being used as a line of advance for an infantry unit.

current U.S. practice, including particularly who is conducting attacks and where they are taking place.

As previously noted, attacks are strictly limited to military objects, defined in Article 52 of Additional Geneva Protocol I of 1977 (AP I) as "those objects which by their nature, location, purpose or use make an effective contribution to military action and whose total or partial destruction, capture or neutralization, in the circumstances ruling at the time, offers a definite military advantage."[*] The law recognizes that in many cases civilian casualties will be unavoidable during strikes against otherwise lawful targets and adopts the principle of proportionality as the standard for judging their legality. AP I article 57 para. 2. (b) (iii) requires refraining from an attack "expected to cause incidental loss of civilian life, injury to civilians, damage to civilian objects, or a combination thereof, which would be excessive in relation to the concrete and direct military advantage anticipated." The "Rendulic Rule" adopted by post-World War II war crimes trials employs a "reasonable commander" standard for liability in this area – if a reasonable commander would conclude an attack to be lawful based on the information they have or should have access to, then there is no criminal liability if it results in disproportionate civilian casualties. Civilian losses have clearly been a major political issue in U.S. strikes in Afghanistan and Pakistan but without detailed information on the value of the intended targets, facts surrounding the decision process, and accurate information about the resulting "collateral" casualties, it is impossible to remotely judge the lawfulness of individual strikes. Nevertheless publicly available information suggests this is an area requiring significant attention on an ongoing basis for both legal and policy reasons.

A unique aspect of the current drone strikes that seemingly differentiates them from most historic uses of force is the apparent deliberate long range targeting of specific individuals or small groups rather than physical objects of military value. There is nothing inherently problematic about selective targeting provided that the selected individuals are otherwise lawful objects of attack. A fundamental principle of armed conflict is that the adversary's combatants: members of the armed forces and affiliated groups and units other than medical and religious personnel, are liable to attack at virtually any time or place during the duration of hostilities. Despite discussions of "battlefields" by some commentators as if it was a limitation on the scope of an armed conflict, the term is merely descriptive, not legal. A battlefield is nothing more than a location where a conflict party finds an adversary or military objective and conducts an attack. Once an individual affiliates with the armed forces, they become a lawful object of attack whether engaged in battle, on the playing fields of West Point, or even at home on leave with their family—subject only to the rule of proportionality governing civilian losses. Citizenship is also irrelevant under the law of war – it is the affiliation with the enemy's forces rather than nationality per se that renders an individual liable to attack. The Supreme Court noted in *Ex parte Quirin* that U.S. citizenship was no bar to detention and trial of an individual as enemy who had violated the law of war, and the inherent logic would have supported engaging them in combat had they not been captured.

A complicating factor in the current conflict is the United States' failure to clearly classify our adversaries within any recognized law of war categorization. If we consider al Qaeda and

---

[*] Although the United States has refused to ratify Additional Protocol I, this provision, along with many other parts of the treaty are recognized as being declaratory of customary international law and accepted as binding on all states.

Taliban fighters as combatants then we can lawfully kill them or detain them for the duration of hostilities based simply on establishing that status. The fundamental privilege that the law of war confers on a combatant in exchange for this vulnerability is immunity from domestic laws, which ordinarily criminalize any act of violence to persons or property. As a result of this immunity, sometimes called "the combatant's privilege," their conduct must be judged under the law of war rather than ordinary criminal laws. We have refused, however, to accord members of al Qaeda and the Taliban the basic right to engage in combat against us. We have instead treated any such conduct, such as Omar Khadr's alleged throwing a grenade at an attacking U.S. soldier, as criminal on the ground that these are not uniformed military personnel legally entitled to engage in hostilities. As a matter of law, this is tantamount to declaring these adversaries to be civilians. Civilians who engage in hostile activity can still be attacked, but only for such time as they are directly participating in hostilities. This classification thus imposes additional limitations on our authority to conduct drone strikes (or any other attacks) against them. There have been suggestions that U.S. targeting may have been expanded, at least for some period of time, to include Afghan drug traffickers who were supporting the Taliban with sale proceeds. This would clearly be unlawful by law of war standards, as would direct attacks on other individuals who are merely performing non-combat support functions, such as financiers, bookkeepers, propagandists, etc.

This issue is equally relevant to who conducts attacks on our behalf. There is no question that uniformed military personnel, whether regular, reserve, or national guard in federal service are lawful combatants entitled to "fly" drone strikes in a recognized armed conflict. But CIA personnel are civilians, not combatants, and do not enjoy any legal right to participate in hostilities on our behalf. It is my opinion, as well as that of most other law of war scholars I know, that those who participate in hostilities without the combatant's privilege do not violate the law of war by doing so, they simply gain no immunity from domestic laws. Under this view CIA drone pilots are liable to prosecution under the law of any jurisdiction where attacks occur for any injuries, deaths, or property damage they cause. But under the legal theories adopted by our government in prosecuting Guantánamo detainees, these CIA officers as well as any higher level government officials who have authorized or directed their attacks are committing war crimes.

The final issue is one of geography. While lawful attacks are limited by specific concepts of "battlefields," they are nevertheless geographically limited by traditional international law rules to the territory of states party to the conflict as well as international waters and airstrikes. Armed attacks may only be carried out in the territory of a "neutral" state either with that state's permission or if the neutral fails to prevent opposing forces from using its territory to the detriment of another belligerent. The 1970 Cambodia "incursion" is perhaps the paradigmatic example of this principle. Although it provoked a huge domestic outcry, including the protest which resulted in the Kent State killings, and may therefore have been a political mistake, there is little real doubt that it was legal. International rules, which the United States government played the leading role in developing in the wake of the 1837 Caroline incident involving a British military incursion into New York, make it clear that the use of military force on the territory of a neutral without its consent are limited to situations of "the most urgent and extreme necessity" and can only be justified upon showing "a necessity of self-defence, instant, overwhelming, leaving no choice of means, and no moment for deliberation."

It is entirely plausible that the governments of countries where U.S. strikes have taken place, such as Pakistan and Yemen, have in fact given their confidential approval and must simply maintain a public posture of denial for domestic political consumption. But if they have not given their permission, it is hard to believe that many of the strikes conducted realistically satisfy the imminency requirements established by international law. This law is intended, by design, to limit cross-border attacks to only the most extraordinary circumstances.

While some commentators have suggested that the United States can conduct drone attacks in self-defense beyond the scope of the congressionally-authorized armed conflict against al Qaeda and the Taliban, it is important to note that the same stringent legal criteria established for the violation of neutral territory apply to anticipatory self-defense as well. The 2004 report of the United Nations Secretary General's High-level Panel on Threats, Challenges and Change rejected the validity of the Bush Administration's expansive preemptive war doctrine while acknowledging that customary international established a narrow right of anticipatory self-defense even under the U.N. Charter but that it was limited to situations where "the threatened attack is imminent, no other means would deflect it, and the action is proportionate." A sustained, even if sporadic, campaign of drone strikes beyond the scope of hostilities approved by Congress would also raise domestic constitutional issues.

Conclusion

Drone capabilities are progressing rapidly as the United States leads the way in developing and fielding these weapons systems. While the technology is new, it is a mistake to assume that old law is therefore inapplicable. Congress has authorized the President to conduct an armed conflict against al Qaeda and the Taliban, and the law of war includes governing rules and principles broad enough to provide meaningful legal regulation of drone employment. This law authorizes much of what the United States seeks to accomplish with these systems, at least when operated by actual military personnel. But some matters, such as the use of CIA personnel to conduct armed attacks clearly fall outside the scope of permissible conduct and ought to be reconsidered, particularly as the United States seeks to prosecute members of its adversaries for generally similar conduct.

Testimony of William C. Banks

before the Subcommittee on National Security and Foreign Affairs,

Committee on Oversight and Government Reform,

United States House of Representatives

April 28, 2010

Mr. Chairman, Ranking Member Representative Flake, and Members of the Committee. Thank you for the opportunity to testify concerning the legality of unmanned targeting and the use of drones. I direct the Institute for National Security and Counterterrorism (INSCT) at Syracuse University, where I am the Board of Advisers Distinguished Professor of Law and a professor of public administration in the Maxwell School of Citizenship and Public Affairs. I have been engaged in teaching, writing, and speaking about United States national security and counterterrorism law for more than twenty years.

My prepared testimony provides an overview of the law that applies to the use of drones in targeted killing. In my oral remarks, I will focus on the laws of the United States that govern CIA involvement in unmanned targeting.

Introduction

On the first night of the campaign against al Qa`ida and the Taliban in Afghanistan in October 2001, the United States nearly had a major success. Officials believed that they had pinpointed the location of the supreme leader of the Taliban, Mullah Muhammad Omar. While patrolling the roads near Kabul, an unmanned but armed drone trained its crosshairs on Omar in a convoy of cars fleeing the capitol. Under the terms of an agreement, the CIA controllers did not have the authority to order a strike on the target. Likewise, the local Fifth Fleet commander in

Bahrain lacked the requisite authority. Instead, following the agreement they sought approval from United States Central Command (CENTCOM) in Tampa to launch the Hellfire missile from the Predator drone positioned above Omar.

The Predator followed the convoy to a building where Omar and about 100 guards sought cover. Some delay ensued in securing General Tommy R. Franks' approval. One report indicated that a full-scale fighter-bomber assault was requested, and that General Franks declined to approve the request on the basis of legal advice he received on the spot.[1] Another report suggested that the magnitude of the target prompted General Franks to run the targeting by the White House.[2] Media reports indicated that President Bush personally approved the strike, although the delay permitted time for Mullah Omar to change his location and thus disrupt the attack.[3] F-18s later targeted and destroyed the building, but Omar escaped.[4] Some speculated that the attack was aborted because of the possibility that others in a crowded house might be killed.

The decision to target specific individuals with lethal force after September 11 was neither unprecedented nor surprising. In appropriate circumstances the United States has engaged in targeted killing at least since a border war with Mexican bandits in 1916.[5] In a time of war, subjecting individual combatants to lethal force has been a permitted and lawful instrument of waging war successfully. But new elements of the targeted killing policy emerged in recent years, in response to terrorism and its threats against the United States at home and abroad.

The components of the targeted killing policy quickly took on a sharper focus soon after September 11. For the first time, pilotless drone aircraft were equipped both with sophisticated surveillance and targeting technology and with powerful Hellfire missiles capable of inflicting lethal force effectively from a safe distance. After the near miss on Mullah Omar, no verified intelligence reported seeing, much less targeting, either Omar or Usama bin Laden during the Afghanistan campaign. However, on November 3, 2001, a missile-carrying Predator drone killed Mohammed Atef, al Qa`ida's chief of military operations, in a raid near Kabul.[6] Then, in early May 2002 the CIA tried but failed to kill an Afghan factional leader, Gulbuddin Hekmatyar, an Islamic fundamentalist who had vowed to topple the government of Hamid Karzai and to attack U.S. forces.[7]

The calculus for targeted killing changed dramatically on November 3, 2002, when a drone fired a Hellfire missile and killed a senior al Qa`ida leader and five low-level operatives traveling by car in a remote part of the Yemeni desert.[8] In the first use of an armed Predator outside Afghanistan or, indeed, the first military action in the war against terrorism outside Afghanistan, Qaed Salim Sinan al-Harethi was killed. Al Harethi was described as the senior al Qa`ida official in Yemen, one of the top ten to twelve al Qa`ida operatives in the world, and a suspect in the October 2000 suicide bombing of the U.S. destroyer Cole, where 17 American Navy personnel were killed. U.S. intelligence and law enforcement officials had been tracking his movements for months before the attack. Along with al Harethi, killed in the Predator strike were five other al Qa`ida operatives, including an American citizen of Yemeni descent, Kamal Derwish, who grew up in the Buffalo suburb of Tonawanda and who, according to FBI intelligence, recruited American Muslims to attend al Qa`ida training camps.

Now, after years of fighting in Iraq and Afghanistan, the focus of attention for lethal targeting has shifted to Waziristan and neighboring border areas in Pakistan. Last August 5, Baitallah Mehsud was killed when two Hellfire missiles were fired from a Predator drone piloted by someone at CIA headquarters in Virginia. Mehsud was the commander of the Pakistan Taliban. He had terrorized the Pakistani government for years, kidnapped Pakistani soldiers, deployed suicide bombers into the streets of Pakistan, masterminded the assassination of Prime Minister Benazir Bhutto, and was implicated in attacks on U.S. forces in Afghanistan. The missiles struck while Mehsud was lying on the roof of his father-in-law's house, apparently while receiving an intravenous drip for his diabetes or a kidney condition. His wife and uncle were killed, along with his in-laws and eight others, including Mehsud's bodyguards. To many, news of Mehsud's death underscored an important victory against terrorists. To some others, his death was murder. It was significant that Mehsud was in Pakistan, not Afghanistan, and that the trigger was pulled by the CIA, not the U.S. military. Was Mehsud a combatant involved in an armed conflict with the United States in Pakistan? Alternately, was he a civilian who was taking a direct part in hostilities during an armed conflict? If there was no armed conflict in Pakistan when Mehsud was targeted, did the United States nonetheless possess the U.S. and/or international legal authority to target Mehsud with lethal force?

Overview of the Law that Applies to the use of Drones in Targeted Killing

During his campaign, President Obama promised to pursue terrorists around the world, including in their refuges in Pakistan. In 2009, President Obama ordered more drone strikes than President Bush ordered in two terms as President. In the first months of 2010, the pace quickened, as more than a dozen strikes were carried out in the first six weeks of the year, killing up to ninety suspected militants. The administration's legal position was outlined by State Department Legal Adviser Harold Koh in a March 25 speech. Koh offered a vigorous defense of the use of force against terrorists, including the targeting of persons "such as high-level al Qaeda leaders who are planning attacks."[9] Koh indicated that each strike is analyzed beforehand based on "considerations specific to each case, including those related to the imminence of the threat, the sovereignty of the states involved, [and] the willingness and ability of those states to suppress the threat the target poses."[10] Koh indicated that the operations conform to "all applicable law,"[11] and are conducted consistent with the principles of distinction and proportionality. Just what constitutes "all applicable law" in the use of drones in targeted killing?

Regardless of the policy efficacy of the drone strikes, it is never sufficient under the rule of law that a government policy is wise. It must also be supported by law, not just an absence of law violations, but positive legal authority. Indeed, where the subject is intentional, premeditated killing by the government, the need for clearly understood legal authority is paramount. After all, legal authority is what distinguishes murder from lawful policy.

Under the Constitution, the President may order targeted killing in defense of the United States in war. The President's authority as Commander in Chief to "repel sudden attacks"[12] has traditionally had a real time dimension, or a sort of imminence requirement, by analogy to the doctrine of self-defense at international law. Yet a terrorist attack is usually over before it can be repelled in real time, and when the attack is a suicide attack, it is impracticable to strike back. In addition, the United States has learned to expect terrorists to pursue a course of continuing attacks against us. As such, over time a domestic law anticipatory self-defense custom has emerged that permits the President to use deadly force against a positively identified terrorist if he has exhausted other means of apprehending him.[13] Congress surely has the legal authority to regulate the use of force in this setting, and it has done so.

The National Security Act of 1947 authorized the CIA to "perform such other functions and duties related to intelligence affecting the national security as the President or National Security Council may direct."[14] Although the original grant of authority in 1947 likely did not contemplate targeted killing, the 1947 Act was designed as dynamic authority to be shaped by practice and necessity, and by the 1970s, the practice came to include targeted killing. After the

Church Committee learned of and disapproved assassination plots of the CIA or its agents in the mid-1970s, President Ford issued an executive order prohibiting CIA involvement in assassination (but notably not restricting targeted killing) and Congress enacted intelligence oversight legislation that, as amended, continues to require reporting to Congress by the President of significant anticipated intelligence operations.

In the weeks after September 11, President Bush signed an intelligence finding giving the CIA broad authority to pursue terrorism around the world.[15] A finding contains the factual and policy predicates for the intelligence activities authorized in any significant operation, and the document must be personally approved by the President. By statute, a finding must accompany any covert operation approved by the President, including those that permit targeted killing. (The military use operations orders, and are thus neither given authority nor restricted by the findings.) In the classified finding, the President delegated targeting and operational authority to senior civilian and military officials. Revised findings, including any prepared by President Obama, along with their precise approval mechanisms, remain classified. The authority given in these presidential findings is surely the most sweeping and most lethal since the founding of the CIA. In part, the findings contemplate a high and unprecedented degree of cooperation between the CIA and Special Forces, as well as other military units.

Terrorists were first singled out by name in a 1995 Executive Order by President Clinton that introduced a category of "specially designated terrorists" on a list maintained by the Secretary of State and the Treasury Office of Foreign Assets Control.[16] In fact, the CIA has been authorized since 1998 to use covert means to disrupt and preempt terrorist operations planned by Usama bin Laden. The Clinton administration directive was affirmed by President Bush before September 11 and was based on evidence linking al Qa`ida to the August 1998 bombings of U.S.

embassies in Africa. The directive stopped short of authorizing targeted killing, but did authorize lethal force for self-defense.

The 2001 finding was apparently modified in 2006 by President Bush to broaden the class of potential targets beyond UBL and his close circle, and also extends the boundaries beyond Afghanistan.[17] In permitting explicitly the targeting of an individual with lethal force, the finding also more narrowly focuses the potential to inflict violence. Because the Yemen strike was authorized by the President in an intelligence finding, at first blush, the relevant law is the law of intelligence. Since the Hughes-Ryan Amendment of 1974,[18] Congress has authorized CIA covert operations if findings are prepared and delivered to select members of Congress before the operation described, or in a "timely fashion" thereafter. So long as the intelligence committees are kept "fully and currently informed," the intelligence laws permit the President broad discretion to utilize the nation's intelligence agencies to carry out national security operations, implicitly including targeted killing.[19] Such an operation would follow intelligence law as an "operation in foreign countries, other than activities intended solely for obtaining necessary intelligence,"[20] and thus presumably would be conducted pursuant to statutory authority.

To some it seemed that the 2001 finding ran counter to the long-standing ban on political assassination. Enshrined in an executive order first by President Gerald Ford and unchanged since President Reagan's iteration in 1981, the directive forbids political assassination but does not define the term.[21] Just what does distinguish lawful targeted killing from unlawful political assassination? The answer turns upon which legal framework applies. During war, whether authorized by Congress or fought defensively by the President on the basis of his authority, targeted killing of individual combatants is lawful, although killing by treacherous means— through the use of deceit or trickery—is not. In peacetime, any extra-judicial killing by a government agent is lawful only if taken in self-defense or in defense of others. But what rules apply when the United States is engaged in a nontraditional war on terrorism, or war against al Qa'ida? The evolving customary law of anticipatory self-defense and intelligence legislation regulating the activities of the CIA supply adequate, albeit not well articulated or understood legal authority for the drone strikes.

In addition to the President's constitutional authorities as commander in chief and his authorities over intelligence activities authorized by statute, the President's finding may also be supported by Congress's September 14, 2001 Authorization for the Use of Military Force (AUMF) giving the President the authority to use "all necessary and appropriate force" against "persons he determines planned, authorized, committed, or aided the terrorist attacks" of September 11. The sweeping authority granted in the resolution is not time-limited; nor does it have a geographic constraint. Nor is his discretion on choice of target narrowed in any way, so long as the target is connected to September 11 and al Qa'ida.[22] In my view, Congress should revisit the AUMF, now nearly nine years after enactment, and provide a more fine-grained authorization for the use of military force against terrorists. Criteria should be supplied for the use of force in self-defense, including targeted killings, within and outside what are regarded as traditional battlefields. Congress should debate the criteria for triggering the use of lethal force against suspected terrorists. Is functional membership in al Qa'ida or a related group sufficient? Must the target be taking a direct part in hostilities, or is providing financial or logistical support to terrorists enough to permit targeting with lethal force? To what extent should the consent of a sovereign state be required before military force is used against terrorists who seek refuge in that state?

Under what conditions could a U.S. citizen be subject to a Predator attack, ordered by the CIA or the military? Before September 11, the government's authority to kill a citizen outside the judicial process was generally restricted to situations where the American is threatening directly the lives of other Americans or their allies.[23] Still, the President's intelligence finding does not make any exception for Americans. The authority to target U.S. citizens is thus implicit, not explicit. In addition, if the AUMF authorizes strikes against al Qa'ida operatives, there may be authority to use lethal force against the radical American-citizen cleric Anwar al-Awalki, who is apparently hiding in Yemen, and who has shifted from encouraging attacks on the United States to directly participating in them.

The defensive use of force—targeted at a known al Qa`ida leader, for example—also has firm legal roots in customary international law. In making operational decisions like the one made to strike with the Predator in Afghanistan, Yemen, or Pakistan, the international and U.S. law concerning self-defense permits targeting al Qa`ida combatants, although carrying out the strike in a terrorist sanctuary (Pakistan or Yemen, for example) rather than on a traditional battlefield complicates the international legal issues.

On the one hand, President Bush asserted forcefully that the September 11 attacks were acts of war directed at the United States, giving it the legal right to repel the horrific attacks. Secretary of Defense Donald Rumsfeld opined, "it is certainly within the president's power to direct that, in our self-defense, we take this battle to the terrorists and that means to the leadership and command and control capabilities of terrorist networks."[24] Whether waged against us by a state or a non-state terrorist organization, war is defined by what it does, not by the identity of the perpetrator. Still, the law of armed conflict has not yet evolved to account adequately for the twilight zone between conventional war and conventional peace, when nations are subject to the continuing threat of terrorist attack.

On the other hand, within this twilight zone of threat from terrorist attacks it is not clear exactly what distinguishes a combatant and, thus, a proper target, from a civilian who may not be targeted. Nor is it known what evidence will suffice that someone who does not wear a uniform and who does not fight for a sovereign state is sufficiently implicated in terrorist activities so as to warrant targeting with lethal force. Clearly someone who is positively identified as an al Qa`ida operative is an enemy combatant, one who may be targeted with lethal force.

Under international humanitarian law, during an armed conflict the selection of individuals for targeted lethal force is lawful if the targets are combatant forces of another nation, a guerilla force, or a terrorist or other organization whose actions pose a threat to the security of the United States. The United States is surely engaged in an armed conflict in Afghanistan, but the absence of sustained fighting over a significant period of time in Yemen means that there is no armed conflict going on there. Pakistan is a closer case. Until sometime in 2009, and the combined campaign of the Pakistani military and the stepped up use of drone attacks by the Obama administration, the conditions in the border region of Pakistan did not likely rise to the level of an armed conflict under the laws of war. By now, however, the conditions on the ground there have changed so dramatically that the laws of armed conflict may apply in Pakistan, in relation to the United States and its use of military force against al Qa'ida or Taliban insurgents.

Conclusion

Contemporary laws have not kept up with changes in the dynamics of military conflicts. Nowhere is the weakness of the legal regime more glaring than in its treatment of targeted killing. The relevant spheres of authority overlap – the laws of the United States (constitutional, statutory, executive, and customary), international laws (treaty-based and customary), and international humanitarian law (a subset of international law that applies during "armed conflicts"). The relationship of the spheres of authority to one another, and their application as binding law is fraught with dispute and contentiousness. In part, the lack of consensus on the legal rules reflects the changing nature of asymmetric warfare. The United States now finds itself engaged in military conflicts with non-state groups, and such conflicts were not the subject of the extensive international framework for warfare negotiated after the World Wars.

These new battlefields require adaptations of old laws, domestic and international laws. My testimony has shown how the legal authority to permit and regulate targeted killing may be found within the existing legal corpus. Admittedly, however, the foundational authorities are not well formed, and there has been little deliberative attention to modernizing the law to reflect the modern battlefield. Congress would do all of us an important favor by devoting attention to articulating policy and legal criteria for the use of force against non-state terrorists.

---

[1] See Seymour Hersh, "King's Ransom," *The New Yorker* (October 10, 2001). Available at http://www.newyorker.com/fact/content/articles/011022fa_FACT1.

[2] Michael R. Gordon and Tim Weiner, "A Nation Challenged: The Strategy," *New York Times* (October 16, 2001), A1.

[3] Eric Schmitt, "U.S. Would Use Drones to Attack Iraqi Targets," *New York Times* (November 6, 2002), A16.

[4] Gordon and Weiner, "A Nation Challenged"

[5] See William C. Banks and Peter Raven-Hansen, "Targeted Killing and Assassination: The U.S. Legal Framework," 37 U. Richmond L. Rev. 667, 688 (2003).

[6] James Risen, "A Nation Challenged: The Terror Network," *New York Times* (December 13, 2001) at A1.

[7] Thom Shanker and Carlotta Gall, "U.S. Attack on Warlord Aims to Help Interim Leader," *New York Times* (May 9, 2002). Available at http://query.nytimes.com/search/article-page.html?res=9D04E2D91330F93AA35756C0A9649C8B63.

[8] John J. Lumpkin, "*Al-Qaida Suspects Die in U.S. Missile Strike*," Associated Press (November 5, 2002). Available at http://www.timesunion.com/AspStories/story.asp?storyID=68947.

[9] "The Obama Administration and International Law," Keynote Speech at the Annual Meeting of the American Society of International Law, Harold Hongju Koh, Legal Adviser, United States Department of State, March 24, 2010, http://www.state.gov/s/l/releases/remarks/139119.htm

[10] Ibid.

[11] Ibid.

[12] 2 THE RECORDS OF THE FEDERAL CONVENTION OF 1787, at 318 (Max Farrand ed., rev. ed. 1937).

[13] Banks and Raven-Hansen, 37 Richmond L. Rev. at 677-681.

[14] Pub. L. No. 80-253, §102(d)(5), codified as amended at 50 U.S.C. §403-4a(d)(4).

[15] James Risen and David Johnston, "Bush Has Widened Authority of C.I.A. to Kill Terrorists," *New York Times*, (December 15, 2002), available at http://www.nytimes.com/2002/12/15/international.

[16] Gellman, "CIA Weighs 'Targeted Killing' Missions," *Washington Post* (October 28, 2001), A01. Available at http://www.washingtonpost.com/wp-dyn/articles/A63203-2001Oct27.html

[17] David Johnston and David E. Sanger, "Fatal Strike in Yemen Was Based on Rules Set Out by Bush," *New York Times* (November 6, 2002), A16.

[18] "No funds appropriated under the authority of this or any other Act may be expended by or on behalf of the Central Intelligence Agency for operations in foreign countries, other than activities intended solely for obtaining necessary intelligence, unless and until the President finds that each such operation is important to the national security of the United States and reports, in a timely fashion, a description and scope of such operation to the appropriate committees of the Congress . . . ." Pub. L. No. 93-559, §32, 88 Stat. 1804 (1974). The amendment was a component of reforms in intelligence operations law designed to make U.S. covert operations decisions directly accountable to the decision makers. See Stephen Dycus, William C. Banks, and Peter Raven-Hansen, National Security Law 456-459 (4th ed. 2006).

[19] Banks and Raven-Hansen, "Targeted Killing and Assassination," 713.

[20] Pub. L. No. 93-559, §32, 88 Stat. 1804 (1974).

[21] Gellman, "CIA Weighs 'Targeted Killing' Missions."

[22] Banks and Raven-Hansen, "Targeted Killing and Assassination," text at n.482.

[23] John J. Lumpkin, "U.S. Can Target American Al-Qaida Agents," Associated Press (December 3, 2002). Available at http://story.news.yahoo.com.

[24] Ibid.

U.S. House of Representatives
Committee on Oversight and Government Reform
Subcommittee on National Security and Foreign Affairs

Subcommittee Hearing:
"Drones II"

Wednesday, April 28, 2010
Rayburn House Office Building

Written Testimony Submitted By
Kenneth Anderson
April 26, 2010

Honorable Chairman and Members:

1. **Introduction**

2. My thanks to the Subcommittee, the Chairman and Members for inviting me to return to offer further testimony on the subject of unmanned aerial vehicles (UAVs) and "drone warfare."

3. My name is Kenneth Anderson. I am a professor of law at Washington College of Law, American University, Washington DC, and a member of the Hoover Task Force on National Security and Law, The Hoover Institution, Stanford University, Stanford CA. My areas of specialty include the laws of war and armed conflict, international law, and national security law. (A brief biography is attached to this statement.)

4. This testimony follows on earlier testimony that I offered to this Subcommittee on March 23, 2010, on the general strategic and legal issues surrounding the use of drones by both the US military and the CIA in counterterrorism operations worldwide. That earlier testimony, which I do not repeat here, is available at the Subcommittee website.[1] (In addition, two additional background documents on this topic – an article in the Weekly Standard, "Predators Over Pakistan," and a policy chapter in a Brookings Institution volume, "Targeted Killing in US Counterterrorism Strategy and Law," are also downloadable at SSRN.com.)[2]

---

[1] The March 23, 2010 KA testimony is also available as a pdf file downloadable from SSRN.com, at http://papers.ssrn.com/sol3/papers.cfm?abstract_id=1579411.
[2] Kenneth Anderson, "Predators Over Pakistan," Weekly Standard, March 8, 2010, available at http://papers.ssrn.com/sol3/papers.cfm?abstract_id=1561229. Kenneth Anderson, "Targeted Killing in US Counterterrorism Strategy and Law," in Benjamin Wittes, ed., Legislating the War on Terror: An Agenda for Reform (2009 Brookings Institution Press), available at http://papers.ssrn.com/sol3/papers.cfm?abstract_id=1415070. In addition, I have blogged extensively on this topic at the international law professor blog, Opinio Juris, www.opiniojuris.org, and at the law professor blog, Volokh Conspiracy, www.volokh.com.

5. DOS Legal Adviser Harold Koh's March 25, 2010 Statement

6. Harold Koh, Legal Adviser to the State Department, delivered an important speech on international law at the American Society for International Law on March 25, 2010, in which, for the first time, a senior lawyer – indeed, the most senior lawyer on international law – delivered a defense of the lawfulness of the US use of drone warfare.

7. As someone who has been a sharp critic of the silence of the administration's senior lawyers on this crucial topic (indeed, in my earlier testimony before this Subcommittee, on March 23), I welcome and applaud the Legal Adviser's forthright and robust statement. Although relatively short and, as the Legal Adviser noted, an authoritative and considered statement of US legal views rather than a formal legal opinion, it went a long way to assuaging concerns I had expressed on whether the administration's lawyers would stand with the political leadership on so important an issue.

8. The Legal Adviser's statement was noteworthy on several grounds. They include particularly the unmistakable and repeated distinction drawn between "armed conflict" in a strict legal sense and "self defense" as a separate basis for the use of force by the United States. In addition, the statement defended the development of such technologies and the efforts by the United States to develop technologically more sophisticated means of targeting and reducing collateral damage. It stated that there was no obligation to provide targets with "process" prior to striking. And it specifically stated that whether in armed conflict or self-defense, such operations did not violate the US domestic ban on "assassination."

9. The comprehensiveness of this defense of drone warfare, on both armed conflict and self-defense grounds, is impressive and persuasive, and is a major statement of the US view of international law on this topic.

10. The fact that this statement was delivered by the Legal Adviser to the State Department sends an important signal that this is not simply the view of DOD, or the CIA, or any intelligence agency – but the view of the United States, expressed as its "opinio juris," its considered view of its own obligations under international law, to the rest of the world.

11. The Distinction Between "Armed Conflict" and "Self-Defense"

12. As I testified on March 23, the fundamental question of drone warfare is not really the technological platform, but instead where and who operates it. On a traditional battlefield in the hands of the military, drones are simply another air support platform; issues surrounding use, targeting, and collateral damage are no different than for any other weapons system. The debate arises on two axes, one related to self-defense, and the other related to the role of the CIA.

13. Is it lawful to use drones in uses of force that do not constitute "armed conflict" with a non-state actor (Al Qaeda and similar groups) in a technical legal sense, because where the drone strike might take place is far away from the current places of hostilities? That question was answered firmly in the affirmative by the Legal Adviser's statement; such strikes can be justified, even though separate and distinct from "armed conflict," as lawful self-defense.

14. Such self-defense operations are not governed by the full panoply of treaty laws that attach to armed conflict – neither the full range of armed conflict law that applies conflicts between states, nor the limited Common Article 3 rules that apply to conflict with a non-state actor. Strikes outside of armed conflict can be undertaken under the doctrine of self-defense – and although lacking all the technical rules that only make sense in overt, army-against-army fighting, such self-defense operations must still adhere to fundamental customary law including, as the Legal Adviser noted, the principles of distinction and proportionality that are also foundations of the technical law of armed conflict. In other words, to say that these operations are not governed by treaty law applicable to overt warfare is not to say that they lack legal standards. On the contrary, they must adhere to the customary standards of necessity, distinction, and proportionality.

15. The Legal Adviser's statement reaffirms and reinvigorates the traditional US view of self-defense, and I find little if anything in his statement that is inconsistent with – indeed, in my view it is a clear reaffirmation – of a speech in 1989 by then Legal Adviser Abraham Sofaer on the topic of transnational terrorism and self-defense.

16. **The Lawful Role of the CIA**

17. As I noted in my March 23 testimony, if one crucial issue about drone warfare is where it takes place – leading to the importance of the armed conflict/self-defense distinction – a second crucial issue is who is lawfully empowered to carry it out. In other words, the fault line in the argument over drone warfare is much less the weapon system than who uses it and where. In the hands of the military on the ordinary battlefield, it is not very different from other air platforms. The question, of course, is off the ordinary battlefield and in the hands of the CIA.

18. The lawfulness of the CIA's operations under US domestic law is not at issue. The agency has been tasked by direct orders of the President, under the authority of a complex statute that provides for oversight and accountability within and between the political branches. The fundamental challenges come from influential parts of the "international law community" – NGOs, international organizations, activists, academics, UN officials, and others – who view the use of targeted killing as unlawful under international law, or likely so, and particularly so by the CIA and outside of the technical scope of armed conflict.

19. Drone warfare becomes controversial, in other words, almost entirely when it is used by the CIA – and in places outside of the Afghanistan, or a narrow slice of cross border regions with Pakistan. Used as a weapon in counterinsurgency by the US military, it is just another weapon. Used as a weapon in counterterrorism, however, directed against Al Qaeda and Taliban leadership away from the active locales of hostilities, whether further afield in Pakistan or still further afield in Yemen, Somalia, or elsewhere, by the civilian CIA – that is where the sharp arguments mostly take place.

20. The Legal Adviser nowhere mentions the CIA by name in his defense of drone operations. It is, of course, what is plainly intended when speaking of self-defense separate from armed conflict. One understands the hesitation of senior lawyers to name the CIA's use of drones as lawful when the official position of the US government, despite everything, is still not to confirm or deny the CIA's operations.

21. In my view, the next step in this evolving legal process should be to affirm what the Legal Adviser's statement says without naming the CIA. That process might involve the US government, in a no doubt difficult interagency consultation, in establishing new mechanisms for acknowledging operations that are widely known, widely discussed even by senior officials – if not operationally, then at least for the "hypothetical" of asserting their legality. In my view, it would be best to establish a formal category of unacknowledged but also obviously not covert operations by the CIA, with their own mechanisms of Congressional oversight and accountability.

22. The Legal Adviser's statement has announced the framework and done everything but say the words "CIA." It is time to take the next step and say "the CIA." The officers of the CIA who carry out these operations, whether planning or execution, merit the public acknowledgment that what they do is legal.

23. Congress ought to make clear, through pronouncement and resolutions and, even better, through legislation, that any attempts to use international or foreign legal process to go after these officers in pursuit of their duties at the intelligence agencies would be regarded as a serious and unfriendly act toward the United States. It is crucial that the two political branches send a single message that the United States stands behind its self-defense operations as such.

24. **Distinctions in the CIA's Roles in Drone Warfare**

25. The distinction between operations in armed conflict, as a technical legal status, and self-defense operations raises a legal issue that has been at the center of some of the criticism of CIA operations. Some commentators, including eminent laws of war scholars, have suggested that the activities of the CIA operating drones (including from locales in the United States) in the context of the armed conflict in AfPak constitutes unlawful combatancy by CIA personnel.

26. The question is not an idle one, if the State Department's position of a distinction between these two grounds for using force is accepted. The legal rules applicable to participants are different as between these two statuses. In an armed conflict, the rules by which combatants must distinguish themselves in order to qualify for combatant immunity apply, and there is a question as to whether that applies to CIA personnel whose activity is part of the armed conflict underway in the Afghanistan or Pakistan theatres.

27. I do not propose to offer here a detailed or definitive answer to the complex question of when and who is required, for example, to wear uniforms and what those must be in order to meet the requirements of the law of armed conflict. My understanding of the DOD view with regards to its own special forces in the early stages of the Afghanistan war, for example, was that context mattered and that US special forces, while commingled with Northern Alliance fighters, could dress as those militia fighters did, and meet the requirements. Context matters, including personnel flying a drone from an office in the United States.

28. In addition to the question of uniforms and marks identifying combatants, these questions also raise important questions for the United States, and DOD particularly, as to the lawful role of civilians in armed conflict. It is partly a question of the lawful role of civilians in support of US combatant forces, as well as when persons on the other side become lawful targets. The underlying question is the much-vexed topic in international law of war over what constitutes "direct participation in hostilities" (DPH). The views that most matter on this are those of the DOD, and I believe it is premature to opine on them until DOD has expressed an official view.

29. Similarly controversial and important is what constitutes the "armed conflict" in AfPak – does it cover all of Pakistan, so that these questions of DPH by CIA personnel matter as a matter of armed conflict law? Or are parts of Pakistan outside of the active border zones of overt hostilities – such that CIA operations there take place instead under the rationale of "self-defense," and so outside of the technical rules of armed conflict, including the formal rules for combatancy? I do not have definitive views on these questions, but I believe they are appropriate to put to the administration, DOD as well as DOS and the CIA, and to encourage the executive to express a view that it believes can serve as a long-run basis for US practice and legal policy.

30. That said, I do not believe that the CIA's current participation either in the theatre of hostilities or elsewhere is unlawful or contrary to international law under the laws of war or otherwise. There might, however, be useful prophylactic measures that could be taken to make that evident.

31. I believe it is appropriate for Congress to ask that the administration undertake a process to formulate a view of the participation of CIA personnel in an armed

conflict and what that means – a process that would likely be difficult because of issues of interagency review. However, in order to put the CIA's activities on the best long-run legal footing, I believe that Congress should urge the administration to undertake what might be an arduous process of consultation and formulation of views.

32. **Why Ever the CIA? Why Not *Always* the US Military to Use Force?**

33. Lurking just behind many of the questions about the lawfulness of the CIA's use of drones, where and how, is a much bigger policy issue. Congress ought to address it forthrightly. Why should the CIA, or any other civilian agency, ever use force (leaving aside conventional law enforcement)? Even granting the existence of self-defense as a legal category, why ever have force used by anyone other than the uniformed military? Drones raise this question if for no other reason that they can be operated far away from the strike zone, whether by CIA or military personnel. Why ever have the CIA use force?

34. That question is in some sense beyond what a hearing such as this can answer, but I raise it because one way or another, it needs to be addressed. It is behind many of the criticisms that are perennial in this activity. For this testimony, suffice it to say that the United States and many, many other leading countries in the world have found that there are circumstances that both justify the use of covert force, and that serious judgments as to the avoidance of greater harms is best served not by overt military force, but by covert or clandestine force, using civilians.

35. Those judgments might in any instance turn out to be wrong, but as a matter of international practice, states have both possessed and utilized clandestine civilian agents, without acknowledging them, in the past and today. The CIA offers to the President the ability to undertake operations of self-defense on a covert basis that this country – and likewise its friends and enemies alike – has deemed essential since the founding of the CIA at least. I raise this because it is important to recognize that not infrequently, arguments against drones are really proxy for arguments against the very idea of the CIA using force. That is an important argument to have, perhaps, but better to have it on its own terms, not indirectly by arguing about drones.

36. Finally, to be clear, the use of drones, or other use of force by civilian CIA agents in covert operations is not contrary to international law insofar as it is an exercise of lawful self-defense. The traditional view of the United States, as expressed in 1989 and reaffirmed today, is sovereignty and territorial integrity are very important in international law, but they cannot be used to shield transnational terrorists who have found safe haven.

37. **Targeting US Citizens**

38. Press accounts that the Obama administration had affirmatively placed the radical cleric, Anwar Al-Awlaki, currently presumed hiding in Yemen with Al Qaeda in the Arabian Peninsula (AQAP), on the kill-or-capture list and, thus, subject to a drone targeted killing attack, raised some excited discussion in the United States. The fundamental question was whether, under international law or US domestic law, the US government owed a targeted individual some form of judicial process before targeting him or her, or whether perhaps it was unlawful to target the person at all in favor of attempting to arrest and bring for trial under a law enforcement model of counterterrorism.

39. Commentators have correctly noted that that Legal Adviser's statement on drones covers not just targeted killing but, once seen in the context of a US citizen on the kill-or-capture list, is equally a statement that even a US citizen who has joined forces with transnational terrorists and is hiding abroad might be subject to targeted killing. The existing domestic law for making such a determination, the Legal Adviser's statement implies, is sufficient, and there is no obligation in US or international law to provide other process, such as judicial review, before a possible strike. Moreover, this is not an act of assassination, within the meaning of the US domestic prohibition, according to the Legal Adviser's statement. All of this seems to me correct as a matter of law and policy.

40. It bears noting, however, that the Legal Adviser's distinction between armed conflict and self-defense is equally relevant here as in other contexts. That is, it is not necessary that, for example, Al-Awlaki, be a "combatant" to be subject to a strike by the CIA. The legal justification of self-defense is separately available as a basis to attack. The standard in that case is not "combatancy," but the threat posed, immanence, and other traditional factors – including the US's long embrace of "active self-defense," meaning that a threat can be assessed on the basis of a pattern of activity already established in the past, without having to wait until a target is on the verge of acting.

41. Many of the critics of this policy in the United States seem not to appreciate that there is a distinction between the territorial United States, in which the CIA is not authorized to act, and extraterritorially.

42. Critics also frequently raise the spectre that all this is license for the CIA to attack an American – or frankly anyone – in, for example, London or Paris. As stated in my March 23 testimony, however, what is justified in the ungoverned regions of Somalia or Yemen is a different matter applied to places under the rule of law such as our friends and allies. The United States is not going to undertake a targeted killing in London. The diplomatic fiction of the "sovereign equality" of states makes it difficult to say, as a matter of international law that, yes, Yemen is different from France, but of course that is true.

43. The willingness of the Obama administration to assert plainly that it has no trouble targeting an American citizen who has taken up the cause of violence

against the United States and its citizens is a positive sign of resolve in counterterrorism by the administration. For one thing, it lessens at least slightly the incentive of terrorist groups to recruit Americans, which is likely to be no small matter over the long term. But as to the fundamental propriety of targeting an American without judicial process, extraterritorially?

44. The policy is far from unprecedented. Suppose that an American scientist in the Cold War had decided to defect to the Soviet Union with vital nuclear secrets that went to the heart of the US strategic arsenal. US citizen, and not military, and not a combatant, because despite the existence of a Cold War, no actual military conflict was underway. The CIA finds that its best chance to remove the threat is a sniper attack on the US scientist in East Berlin as he attempts to enter the Soviet embassy. Far fetched? Not really. Not an armed conflict and not a "combatant" in a technical legal sense – a US citizen targeted without judicial process, abroad, and under rationales of self-defense. The same concepts apply today, with respect to transnational terrorists, including Americans who take up the cause with them.

45. A Role for Congress?

46. This testimony has highlighted several matters on which Congress could play an important and useful role, primarily in offering support to the policies undertaken by the administration. One of these is to encourage and provide opportunities for other senior lawyers in the intelligence and defense communities to affirm, amplify, and expand on what the Legal Adviser has said.

47. A second role for Congress – and a deeply important one – is to specifically name the CIA as under the protection, so to speak, of these legal views on self-defense. This is of great importance in order to make clear that line officers and legal officials of the CIA are not being put in the untenable position of being tasked to carry out policies for which they might later be accused of violations of international law. Congress needs to clarify its lawfulness, and frankly make clear that countries that seek to gainsay the US's own considered legal view on this topic, including allies and NATO allies, such as unsupervised prosecutors in Spain or elsewhere, will discover that there are consequences.

48. Congress should also invite the administration to elaborate its views of CIA actions inside Pakistan, to state whether it thinks such activities are part of the on-going armed conflict, are separate from it, and how such views interact with legal doctrines of DPH. This is a topic on which there needs to be a coordinated legal view among DOS, DOD, CIA, DNI, and perhaps others.

49. Congress should explicitly endorse the Obama administration's view that American citizenship does not preclude one from being targeted extraterritorially under laws of armed conflict, self-defense or US domestic law.

50. Congress should be strongly supporting, through budget processes and otherwise, the development of more discrete and discriminating drone-and-missile technologies to reduce collateral damage to the minimum that technology can allow, as well as to improve targeting identification.

51. Congress should invite the CIA to share what it believes it can regarding collateral damage involved in drone strikes.

52. If, as some commentators have suggested, the use of force threshold is gradually shifting toward smaller and more discrete uses of force by US actors, made possible by evolving drone technologies, then Congress should revisit the rules regarding oversight, review, and accountability to ensure that they take account of emerging realities about the use of force as perhaps a substitute – and highly desirable substitute – for large scale overt war.

53. Congress should make clear that it rejects utterly the argument made popular in the press in recent months that drone warfare is somehow dishonorable or that it somehow reduces the disincentives for the US to use violence, or that it makes violence too easy for the United States because its forces are not at risk, with the barely concealed implication that if American servicemen and women are not actively at risk of getting killed, because drones make it possible to take the fight to the enemy without having to fight through whole countries on the ground to get there – that drone warfare, that is, is somehow illegitimate, dishonorable, unlawful, or an enabler of the US to let loose its unrestrained propensity to use violence. It is none of those things, and Congress should say so.

54. Conclusion

55. I thank the Subcommittee, chairman and members, for this opportunity testify. Please be in touch with me should you have any further questions or seek additional views.

Kenneth Anderson
April 28, 2010, Washington DC
kanders@wcl.american.edu

Kenneth Anderson
Biography

Kenneth Anderson is a professor of law at Washington College of Law, American University, where he has taught since 1996. He is also a visiting fellow and member of the Hoover Institution Task Force on National Security and Law, Stanford University. Prior to joining the American University faculty, Mr. Anderson was general counsel to the Open Society Institute-Soros Foundations in New York City, and prior to that the director of the Human Rights Watch Arms Division. He is a 1986 graduate of Harvard Law School and 1983 graduate of the University of California, Los Angeles; he clerked in 1986-87 for Justice Joseph R. Grodin of the California Supreme Court. He is a member of the editorial board of the Journal of Terrorism and Political Violence, past Treasurer and Executive Committee member of the Lieber Society of the American Society of International Law, and a blogger at Opinio Juris international law blog and the Volokh Conspiracy law blog. He is the author of numerous articles on international law and laws of war, and served as legal editor of Crimes of War (1998 Norton).

Anthony D. Romero
EXECUTIVE DIRECTOR

April 28, 2010

President Barack Obama
The White House
1600 Pennsylvania Avenue, N.W.
Washington, D.C. 20500

Dear Mr. President:

On behalf of the ACLU and its 500,000 members, I am writing to express our profound concern about recent reports indicating that you have authorized a program that contemplates the killing of suspected terrorists – including U.S. citizens – located far away from zones of actual armed conflict. If accurately described, this program violates international law and, at least insofar as it affects U.S. citizens, it is also unconstitutional.

The U.S. is engaged in non-international armed conflict in Afghanistan and Iraq and the lawfulness of its actions must be judged in that context. The program that you have reportedly authorized appears to envision the use of lethal force not just on the battlefield in Iraq, Afghanistan, or even the Pakistani border regions, but anywhere in the world, including against individuals who may not constitute lawful targets. The entire world is not a war zone, and wartime tactics that may be permitted on the battlefields in Afghanistan and Iraq cannot be deployed anywhere in the world where a terrorism suspect happens to be located. Your administration has eschewed the rhetoric of the "Global War on Terror." You should now disavow the sweeping legal theory that underlies that slogan.

Even in an armed conflict zone, individuals may be targeted only if they take a direct part in hostilities, for such time as they do so, or if they have taken up a continuous combat function. Propagandists, financiers, and other non-combat "supporters" of hostile groups cannot lawfully be targeted with lethal force. Applicable international humanitarian law also prohibits targeted killing except in order to prevent an individual's future participation in hostilities; fighters cannot be targeted solely as retribution for past actions. Furthermore, basic law-of-armed-conflict principles require that in such operations, civilians who are not taking direct part in hostilities must not be targeted, precautions must always be taken to spare the civilian population, anticipated civilian casualties must never be disproportionate to the expected concrete military advantage, and strikes must only occur when required by military necessity.

AMERICAN CIVIL
LIBERTIES UNION
NATIONAL OFFICE
125 BROAD STREET, 18TH FL.
NEW YORK, NY 1004-2400
T/212.549.2500
WWW.ACLU.ORG

OFFICERS AND DIRECTORS
SUSAN N. HERMAN
PRESIDENT

ANTHONY D. ROMERO
EXECUTIVE DIRECTOR

ROBERT B. REMAR
TREASURER

Outside armed conflict zones, the use of lethal force by the United States is strictly limited by international law and, at least in some circumstances, the Constitution. These laws permit lethal force to be used only as a last resort, and only to prevent imminent attacks that are likely to cause death or serious physical injury. According to news reports, the program you have authorized is based on "kill lists" to which names are added, sometimes for months at a time, after a secret internal process. Such a program of long-premeditated and bureaucratized killing is plainly not limited to targeting genuinely imminent threats. Any such program is far more sweeping than the law allows and raises grave constitutional and human rights concerns.

In a series of cases involving prisoners currently held by the U.S. at Guantanamo Bay, your administration has taken the position that the 2001 Authorization for Use of Military Force permits the detention of individuals captured anywhere in the world, even individuals who have no connection to the battlefield. For example, your administration has advanced that argument in the case of one of our clients – Mohammedou Salahi – who was detained in Mauritania. We do not think the AUMF can be read so broadly. In *Hamdi v. Rumsfeld*, the Supreme Court interpreted the AUMF consistently with international law, permitting the detention of a U.S. citizen captured in Afghanistan only because the detention of *battlefield* combatants was "so fundamental and accepted an incident to war as to be an exercise of the 'necessary and appropriate force' Congress has authorized the President to use." 542 U.S. 507, 518 (2004). But even if the AUMF could be read to authorize the *detention* of suspected terrorists apprehended far from any zone of actual combat, it is a far more radical thing to propose that the AUMF authorizes the extrajudicial *execution* of those people. Outside of armed conflict zones, human rights law and the Constitution prescribe strict limits on the use of lethal force, limits that are narrower than those applicable in armed conflicts, and narrower than the standards governing detention. Targeted killing of suspects away from the battlefield is not a "fundamental and accepted ... incident to war." Based on the available information, neither does your targeted killing program appear to be an exercise of "necessary and appropriate force" used only as a last resort to prevent imminent threats. The AUMF may be broad, but the authority it granted was not limitless, and it cannot now be construed to have silently overridden the limits prescribed by international law.

The program you have reportedly endorsed is not simply illegal but also unwise, because how our country responds to the threat of terrorism will in large measure determine the rules that govern *every* nation's conduct in similar contexts. If the United States claims the authority to use lethal force against suspected enemies of the U.S. anywhere in the world – using unmanned drones or other means – then other countries will regard that conduct as justified. The prospect of foreign governments hunting and killing their enemies within our borders or those of our allies is abhorrent.

The program you have endorsed also risks the deaths of innocent people. Over the last eight years, we have seen the government over and over again detain men as "terrorists," only to discover later that the evidence was weak, wrong, or non-existent. Of the many hundreds of individuals previously detained at Guantánamo, the vast majority have been released or are awaiting release. Furthermore, the government has failed to prove the lawfulness of imprisoning individual Guantánamo detainees in 34 of the 48 cases that have been reviewed by the federal courts thus far, even though the government had years to gather and analyze evidence for those cases and had itself determined that those prisoners were detainable. This experience should lead you to reject out of hand a program that would invest the CIA or the U.S. military with the unchecked authority to impose an extrajudicial death sentence on U.S. citizens and others found far from any actual battlefield.

Sincerely,

Anthony D. Romero
Executive Director

www.ingramcontent.com/pod-product-compliance
Lightning Source LLC
Chambersburg PA
CBHW080558090426
42735CB00016B/3279